Electronic Circuits for the Behavioral and Biomedical Sciences

A series of books in PSYCHOLOGY

Editor: Stanley Coopersmith

Electronic Circuits for the Behavioral and Biomedical Sciences

A REFERENCE BOOK OF USEFUL SOLID-STATE CIRCUITS

Mitchell H. Zucker

Electronics Technician, Department of Psychology
University of California, Davis

W. H. FREEMAN AND COMPANY
San Francisco

Printed in the United States of America.

Library of Congress Catalog Card Number: 76-81921

Standard Book Number: 7167 0918-X

Preface

Despite copious publications on the theory and functioning of electronic circuits, the modern researcher in the behavioral and biomedical sciences has had no up-to-date book describing the electronic technology relevant to his particular research needs. Too often the researcher's lack of electronic training has kept him from making adequate adaptations of existing circuitry. The "home-brewed" electronic experiment that turns out to be not only too costly, but also unreliable, is hardly unknown in university research laboratories. This book is written specifically as a reference source for simple, reliable electronic circuits that can be used by the behavioral or biomedical researcher who has had little formal electronic education. It will introduce him to basic semiconductor circuit construction and testing.

The introductory chapters include a discussion of basic semiconductor operation and an explanation of the symbology of circuit diagrams. Particularly important for the novice experimenter are the chapters describing circuit construction techniques and simple, effective troubleshooting procedures. The larger part of the book consists of circuit descriptions, with specific instructions on how to construct them and how to insure that they will operate properly. Three criteria were used in selecting the circuit projects. First, tests must have shown that all the circuits included operate reliably. Second, the circuits must be simple, and demonstrate basic electronic principles. Even though the researcher may have no immediate application for a particular circuit, the description of it may broaden his general electronic understanding. Third, all the circuits must be easily constructed from readily available components. In the few instances where parts for projects are difficult to obtain, addresses have been included to facilitate mail orders. The Appendix includes formulas and tables that may aid in design and construction, and a list of basic relay circuit configurations commonly used in behavioral and biomedical experimentation.

Chapters 5 to 10 constitute a basic inventory of circuit descriptions. By combining various designs, the experimenter can create new circuit configurations suitable for his particular research. For example, should an experimenter need a sensitive photocell detector coupled to an electronic pulse-forming ntework, he could construct such a device after reading the descriptions of the photocell amplifiers and monostable multivibrator. The ability to work creatively with circuits is less likely to come from studying theoretical discussions than from "playing" with actual circuits. The experimenter who modifies circuits and then tries to explain the resulting circuit function in terms of the individual component performance is learning by the best method, by doing.

Although the book was not intended as a textbook, it will be useful for an introductory survey course in electronics, particularly where the intent of the course is to familiarize the student with the circuits and the construction techniques used in electronic apparatus.

I wish to acknowledge the kindness of the General Electric Company and the publishers of *Popular Electronics, Electronics World*, and *Radio Electronics*, who have permitted me to include in this book circuit descriptions and diagrams that originally appeared in their publications. To Vincent Polidora and Henry Veatch I would like to express my gratitude for their penetrating criticisms and helpful corrections. For the many photographs in this book, I am grateful to Ronald R. Peck, Davis, Calif. Various portions of the text have been read and criticized by Allan Wegner and Carl Klein. To both these gentlemen, and the rest of the staff of Night Flight Enterprises, I express my deep thanks and appreciation.

Finally, I wish to especially thank my wife, Rosalie, for her assistance in reviewing and typing the manuscript.

January 1969   Mitchell Zucker
Davis, California

Contents

Electronic Circuits for the Behavioral and Biomedical Sciences

Introduction to
Semiconductor Operation

This chapter is intended for those who want a rudimentary knowledge of semiconductors and how they function. Although the reader need not know the theory of semiconductor operation to build the circuits in this book, he will better understand the circuits he builds if first he has gained some insight into how they work.

The information in this chapter will not enable the experimenter to design his own circuits, for which far more information than can be presented here would be required. However, there are many books and courses of instruction on semiconductor technology for both novice and advanced experimenters. A bibliography listing several of these sources is at the end of this book.

SEMICONDUCTOR MATERIALS

The category of semiconductor devices includes diode rectifiers, silicon-controlled rectifiers, and transistors. These devices are characterized by the materials used in their construction, namely, semiconductor compounds or elements. A semiconductor is any material that displays conductive characteristics intermediate between metals (good electrical conductors) and insulators. Silicon, selenium, and germanium are the elements most commonly used in semiconductor devices. To understand how semiconductor devices are developed and how they function, we must look at the atomic and molecular structure of the semiconducting elements.

An atom of the element germanium, a typical semiconductor used in the manufacture of transistors, has four valence electrons. Atoms of germanium form covalent bonds with other germanium atoms, resulting in a crystal lattice molecular structure. If a very small quantity of arsenic, which has five valence electrons, is added to pure germanium, four of arsenic's valence electrons form covalent bonds with germanium electrons; the fifth arsenic electron is held fairly weakly (see Figure 1.1). The altered germa-

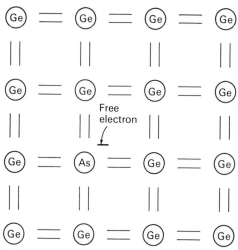

Figure 1.1
Valence electron bonds in germanium, with an arsenic atom replacing a germanium atom.

nium is then said to have a few "free" electrons within its crystal structure. The name given to the germanium crystal containing donor arsenic atoms is N-type (for negative-type) germanium.

If a small amount of a substance having three valence electrons, such as boron or indium, is added to germanium, the resulting crystal structure is that diagrammed in Figure 1.2. The formation of a four-electron bond between an indium atom and germanium crystal creates a "hole" in a germanium atom. The altered germanium crystal is known as P-type (for positive-type) germanium.

When P-type and N-type materials are fused or joined together by a special manufacturing process, a P-N junction is formed. The result is a useful semiconductor device known as a rectifier or diode.

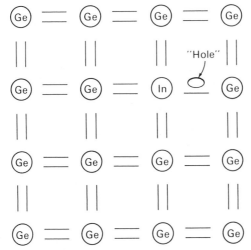

Figure 1.2
Valence electron bonds in germanium, with
an indium atom replacing a germanium atom.
The "hole" indicates an electron absent from
a covalent bond, a site readily occupied
by a free electron.

RECTIFIER OPERATION

Figure 1.3 shows a simple circuit composed of a battery, a resistor, and
a silicon rectifier made of fused N-type and P-type silicon wafers. At the
junction of the two wafers, where the N-type and P-type materials meet,
current can flow from the P region to the N region, but cannot effectively
flow in the reverse direction. (Note that in this and all subsequent discussions, the standard current-flow convention is used; current is described
as flowing from positive to negative.) If the rectifier is reversed (see Figure
1.4), no current can flow, except for a very small current in the microampere range, commonly called leakage current, which for the most part
can be ignored. When a voltage source is connected in this manner, the
rectifier is described as being reverse-biased.

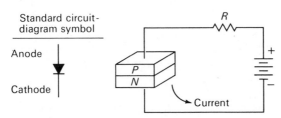

Figure 1.3
Current flow through a silicon rectifier.

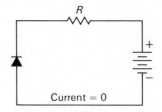

Figure 1.4
No current flow through a
silicon rectifier; reverse bias.

If the rectifier is connected in series with a source of alternating current (see Figure 1.5), the rectifier will conduct electricity only when the positive-going half-cycle is applied to the rectifier's anode or positive lead, since the rectifier is effectively reverse-biased when the negative-going half-cycle

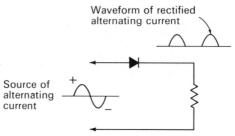

Figure 1.5
Alternating current flow through a rectifier.

is applied to the rectifier anode. In this circuit, the alternating current is said to have been changed to pulsating direct current or to have been half-wave rectified.

A useful circuit using alternating current and a rectifier is shown in Figure 1.6. In this circuit a switch is used to connect either a rectifier in

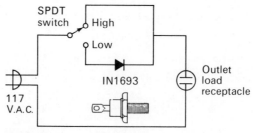

Figure 1.6
High-low switch, showing in outline the IN1693
rectifier diode, oriented in the same direction as
its symbol in the diagram.

series with a load or the load directly to a source of power. In the switch position shown, the power is delivered directly to the output receptacle. When the switch is thrown to its other position, the rectifier is connected in series with the output receptacle and will block half the half-cycles from the source of power. As a result, the load will be presented with only the half-cycles of the line frequency that are passed by the rectifier. Since the total power applied to the load may be expressed as the product of the applied voltage and the current flowing through the load, inserting a rectifier in series with the load reduces power by about half. There is a slight voltage drop across the rectifier when it is conducting in the forward direction, but this drop is quite small compared to the voltage drop in the reverse or blocking direction. By switching from the high position to the low position, one may dim a light, or reduce a heater's output or a motor's speed.[1] With the particular rectifier shown in Figure 1.6, it is possible to control up to 130 watts of power; that is, the rectifier may be placed in series with a 130-watt load without being damaged. Other rectifiers can control loads as great as 1,000 watts.

RECTIFIER OPERATING CHARACTERISTICS

Figure 1.7 is the characteristic curve or operation curve for a typical rectifier. It is plotted by measuring the voltage across the rectifier and the current flowing through it when it is biased in the forward and in the reverse directions. It can be seen in the right half of Figure 1.7 that a forward-biased rectifier has a relatively small forward voltage (V_F) measured across it when conducting a relatively large current (I_F). That is, useful voltage is lost within the rectifier structure and creates heat. The product of V_F and I_F gives the power loss within the rectifier. This power loss and the consequent heating can destroy the rectifier, and must be considered when designing rectifier circuits. This heating is why metal plates, called "heat sinks," are used in some stud-mounted rectifier circuits, since the heat sink helps dissipate heat that would otherwise adversely affect the current rating of the rectifier. The current-handling capabilities of a rectifier are determined by temperature; excess heat must be removed from the rectifier to prevent it from being destroyed.

When the rectifier is reverse-biased (left half of Figure 1.7), a large reverse voltage will cause a relatively small flow of current. There is a certain point, however, at which increases of reverse-bias voltage will cause a very large reverse current flow. The voltage at this point is known as the peak reverse voltage (PRV), and most rectifiers will be destroyed when the PRV is exceeded (Point A in Figure 1.7).

---

[1] Certain loads, such as fluorescent lamp ballasts and transformers, will not operate properly when connected to the rectifier circuit shown in Figure 1.6.

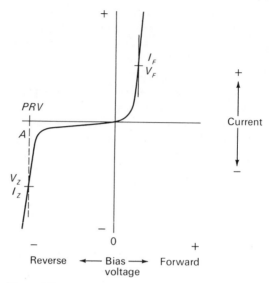

Figure 1.7
Characteristic rectifier curve.

There is a special class of rectifiers, known as zener diodes, which are specifically designed to operate in the reverse region at a voltage known as the zener voltage (V_Z in Figure 1.7). The voltage across a zener diode is nearly constant for any reverse current flow within its operating range. In certain applications the zener diode may be used as a source of reference voltage. In voltage regulators, for example, the load voltage may be held constant by placing the load in parallel with a zener diode. Since the zener diode voltage remains nearly constant under varying load conditions, variations in supply voltage are effectively canceled.

THE TRANSISTOR

By adding a third semiconductor wafer and lead to the two shown in Figure 1.3, a transistor may be constructed. As shown in Figure 1.8, a transistor may have either two P regions and an intermediate N region (known as PNP type) or two N and one P (known as NPN type). Introducing a base lead circuit to the transistor makes conduction possible, as is shown in Figure 1.9 for a PNP-type transistor.

In the base circuit of Figure 1.9, current can flow between the base and emitter leads in the same manner that current can flow from anode to cathode in the rectifier (both pairs are forward-biased). In the collector-emitter circuit, although the current is forward-biased through P-N junction 2, it appears to be reversed-biased through N-P junction 1. However,

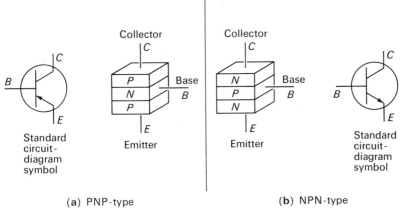

(a) PNP-type   (b) NPN-type

Figure 1.8
Basic transistor configurations and symbols: **(a)** PNP-type; **(b)** NPN-type.

introducing the base circuit causes interaction within the transistor that permits the collector current to flow through the reverse-biased N-P junction 1. In a typical transistor circuit wired with a common emitter, the collector current that can flow is about fifty times the current that can flow in the base circuit. Any increases or decreases in the base current are reflected as increases or decreases in collector current. Thus, small currents flowing in the base circuit can control much larger currents in the collector circuit. This ability of a transistor to control large currents by smaller currents is known as amplification. The ratio of changes in collector current (I_C) to changes in base current (I_B) is known as the base current amplification factor or beta (β). The beta for transistors is usually available in data provided by the manufacturer, and is sometimes specified by the symbol h_{fe}.

In the NPN type of transistor, the current flow is the reverse of that in the PNP type. A typical NPN-type transistor is shown in Figure 1.10. The primary difference between the NPN circuit of Figure 1.10 and the

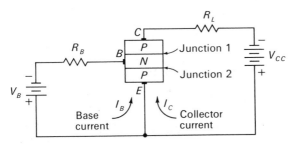

Figure 1.9
Basic transistor circuit, PNP-type.

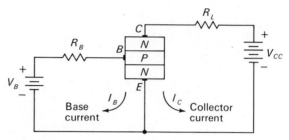

Figure 1.10
Basic transistor circuit, NPN-type.

PNP circuit of Figure 1.9 is that the polarity of the battery connections is reversed. Because of this reversed polarity, the direction of current flow through the collector circuit and the base circuit is reversed also. An analysis of the forward and reverse biases of both the base circuit and the collector circuit will show that current can only flow in the direction indicated in Figure 1.10.

TRANSISTOR BIASING

When the base current (I_B) flowing in the circuit of Figure 1.10 is increased, there is a corresponding increase in collector current (I_C). If the increase in base current is continued, almost all the battery voltage (V_{CC}) will eventually be dropped across the load resistor (R_L). Further increases in base current will have little effect on the collector, whose current flow will be similar to that through a forward-biased diode. This condition of maximum collector-current flow through a transistor is known as "saturation." Conversely, when a transistor is biased so that there is no base current flowing ($I_B = 0$), the transistor is said to be "cut-off" and

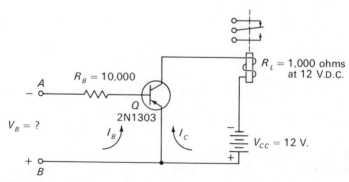

Figure 1.11
Simple electronic relay.

the collector current is at a minimum. By operating the transistor in saturated and cut-off modes, the transistor effectively functions as a two-state
(or binary) electronic switch or relay.

Basic design considerations for biasing in a simple electronic relay circuit
are illustrated in Figure 1.11. Relay R_2 is to operate whenever the necessary
base voltage (V_B) is applied to terminals A and B. To find the minimum
base voltage needed to saturate the transistor, the first step is to determine
the collector current (I_C) needed to operate the relay. This is a simple
Ohm's law equation:

$$I_C = \frac{V_{CC}}{R_L}.$$

$I_C = 12/1,000 = 0.012$ ampere $= 12$ milliamperes.

Next, the base current (I_B) is determined from the beta of the transistor
as found in the data provided by the manufacturer. For the 2N1303, the
minimum gain is 50 ($\beta = h_{fe} = 50$). It follows that

$$\beta = \frac{I_C}{I_B}, \quad I_B = \frac{I_C}{\beta}.$$

$I_B = 0.012/50 = 0.00024 = 0.24$ milliampere.

By again using Ohm's law, the value of V_B can be obtained:

$$V_B = I_B \times R_B$$
$$= 0.00024 \times 10,000$$
$$= 2.4 \text{ volts.}$$

When a minimum voltage of 2.4 volts is applied to terminals A and B,
the transistor will saturate and the relay will operate. This voltage may be
lowered or raised by adjusting the value of R_B. If, for example, the relay
is to be operated with a minimum applied voltage of 6 volts, then

$$R_B = \frac{V_B}{I_B},$$

$R_B = 6/0.00024 = 25,000$ ohms.

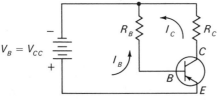

Figure 1.12
Transistor circuit with one
source of voltage.

It is often desirable or necessary to eliminate the two separate sources of voltage for the base circuit and collector circuit. This can be accomplished by using one source of voltage for both circuits (see Figure 1.12). As in the circuit of Figure 1.11, the base circuit and the collector circuit are biased with the same polarity. The base resistor (R_B) is selected to permit a specific amount of base current to flow. The collector current flows through the collector resistor (R_C) and back to the battery; R_C is also known as the load resistor. Since the emitter lead is common to both the base circuit and the collector circuit, the entire circuit is known as a common-emitter configuration, and is the most common circuit configuration used in switching circuits. With this particular configuration, the value of R_B will be high compared with the value of R_C, since the base current flow must be limited more than the collector current flow. The importance of this difference in resistances will be evident if, for resistor R_B, we substitute the resistance of a rat making contact between two grids in a cage, as shown in Figure 1.13.

The resistance of the rat across the grid bars completes the base bias circuit, which allows collector current to flow and activate the relay. Although this circuit appears to fulfill the requirements for a sensitive contact relay, several factors make it inappropriate as a reliable sensing circuit. First, the skin resistance across the paws of a rat, because of dead skin and dirt, might be so high that not enough base current would flow to raise the collector current above the pull-in current required to activate the relay. Second, the rat may urinate across the grid bars. The urine (containing a high concentration of electrolytes, which conduct current very efficiently) would present so low a resistance to the base circuit that the base current might exceed the transistor's rating and damage it. Finally,

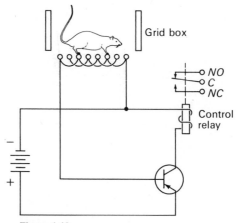

Figure 1.13
Basic contact relay circuit.

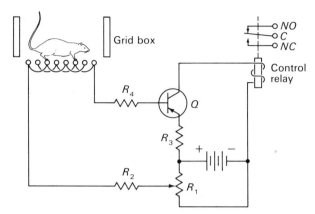

Figure 1.14
Basic contact relay with transistor biasing.

the rat may remain on the grid bars for so long that the relay, being continually energized, might overheat, as might the transistor, enough to be permanently damaged.

The circuit presented in Figure 1.14 shows how transistor biasing can be employed to make the basic contact relay a more appropriate, useful, and reliable circuit. In this circuit a potentiometer (R_1) is added in parallel to the battery. The potentiometer functions as a voltage divider (or sensitivity control), providing a variable source of voltage to the base circuit. By selecting an appropriate base bias voltage, the sensitivity of the contact relay may be adjusted for each operation. For example, if the rat has a relatively high skin resistance, it may be necessary to adjust R_1 so that greater base bias voltage is applied than would be if the rat's skin resistance were relatively low. Resistor R_2 is added to provide a certain minimum

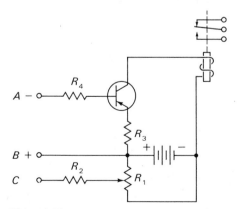

Figure 1.15
Multifunction transistor relay.

amount of resistance to the base circuit. It is used to protect the base circuit from a sudden increase in base current, which might be caused, for example, by the rat urinating across the grid bars. Resistor R_3 is used in the emitter-collector circuit to stabilize the relay operation. Should the rat remain on the grid bars for a long time, the current through the relay circuit would be somewhat reduced by R_3. This reduction helps prevent the relay from overheating and from possibly damaging itself or the transistor; R_3 acts as a temperature-stabilizing element. Resistor R_4 is placed in the base circuit to allow the contact relay to serve an additional function without being damaged. This additional function is illustrated by Figure 1.15, the circuit which is identical to that of Figure 1.14 except for the addition of a terminal (B) emanating from the positive terminal of the battery. If terminal C (R_1 and R_2) is not used, terminals A and B can be used to sense inputs of voltage. An example of how this circuit could be used is shown in Figure 1.16.

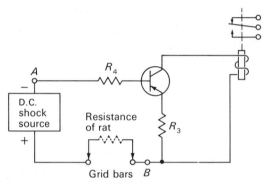

Figure 1.16
Voltage detector or amplifier.

The circuit shown in Figure 1.16 could be used to count the number of times a rat is shocked when it contacts the grid bars. The shock voltage is used to bias the base circuit when the circuit is completed through the rat. The shock voltage functions as in the circuit of Figure 1.9, where two batteries were used to bias both the base circuit and the collector circuit. Resistor R_4 is added to the circuit of Figure 1.16 to prevent excessive currents from flowing through the base circuit. It is also possible to augment R_4 with a rheostat (R_5) connected to the minus side of the battery, in order to provide a sensitivity control for this circuit (see Figure 1.17). By adding some forward bias to the base circuit through the rheostat, the transistor can be made to conduct just below the level necessary to activate the relay. Should the rat's resistance be too high, the added base bias would be provided by the shock source when the rat touches the grid bars.

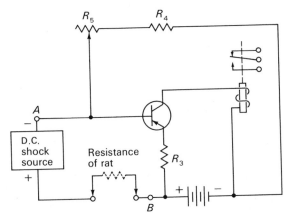

Figure 1.17
Voltage detector or amplifier with sensitivity control.

Resistor R_4 is retained to provide a minimum amount of base bias to the circuit.

The sensitivity of the circuit in Figure 1.17 may be further increased by employing two transistors instead of one. One transistor is used to amplify the input, and its output (collector current) is applied to the base circuit of a second transistor, the collector current of which is applied to the relay coil. The sensitivity of the practical two-stage transistor circuit shown in Figure 1.18 is almost the square of that of a single-stage circuit. This circuit can be made to actuate the relay when the resistance across terminals I and B is as great as 50,000,000 ohms. It can also actuate the relay when currents as low as 0.2 microamperes at approximately 1 volt D.C. are applied across terminals I and G. Note that the transistors used in this

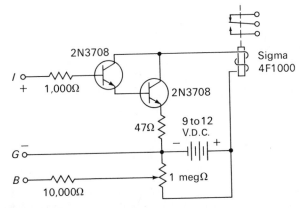

Figure 1.18
Practical sensitive relay circuit.

circuit are NPN types, as opposed to the PNP types used in the circuits of Figures 1.11 through 1.17, and that the battery polarity is therefore reversed; otherwise, circuit operation is identical. (A detailed description of the circuit shown in Figure 1.18 can be found in Chapter 7.)

In addition to biasing the transistor to operate as an electronic switch, it is possible to bias it to some point about midway between cut-off and saturation. When biased in this manner, the transistor can amplify changing signals applied to the base circuit, such as audio- and radio-frequency signals. Although the circuitry needed to amplify changing signals is more complex than the basic amplifier circuit shown in Figure 1.19, the latter will illustrate some of the basic design considerations in constructing an amplifier circuit. A complete discussion of amplifier design would be beyond the scope of this book.

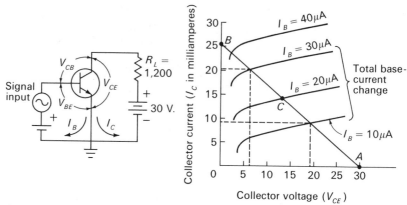

Figure 1.19
Basic transistor amplifier with characteristic curve and load line.

Figure 1.19 shows a simple common-emitter amplifier circuit, and a characteristic curve of the changes in collector current and collector voltage for various values of base current. A similar characteristic curve (without line BCA) is usually provided by the manufacturer in describing the characteristics of a transistor when wired in a common-emitter mode. The line BCA is known as the "load line" and is important in analyzing the operating characteristics of a transistor circuit. It is obtained by first determining the collector voltage (V_{CE}) when the collector current is zero. With the circuit of Figure 1.19, the collector voltage will equal the full 30-volt supply voltage when the collector current is zero. This value is plotted as point A on the characteristic curve. Next, the collector current is determined when the collector voltage is zero. With zero collector voltage the transistor must be fully turned on. The collector current is essentially limited only by the value of R_L and is determined by Ohm's law:

$$I_C = \frac{V_{CE}}{R_L} = 30/1{,}200 = 0.025 \text{ ampere} = 25 \text{ milliamperes.}$$

This point is plotted as point B. A line is then drawn to connect points A and B.

Point C usually represents an operating point around which changes in base current, caused by changes in the signal, produce relatively proportional changes in collector current. In the example in Figure 1.19, it can be seen that within the limits of $I_B = 30$ microamperes and $I_B = 10$ microamperes, changes in base current produce proportional changes in collector current. If the base current is increased to 40 microamperes, the load line intersects the base current line in its nonlinear portion. The resulting collector current will not be proportional and the signal will be distorted. Another source of distortion arises from the fact that higher base-current lines are not equidistant from each other in the linear portions of the characteristic curves. As the base current is raised, the linear portions of the curves become crowded. In designing an amplifier, the load line and operating point are selected to suit the requirements of the circuit.

It can be seen in Figure 1.19 that when the input signals cause total base current changes of 20 microamperes, the collector current will change a total of 12 milliamperes. Since beta (β) is the ratio of change in collector current to change in base current, in this example

$$\beta = \frac{12 \text{ milliamperes}}{20 \text{ microamperes}} = 600.$$

This would indicate a very high-gain transistor circuit.

In addition to the common-emitter circuit configuration described above, other configurations (i.e., the common-base and common-collector) may be constructed, each with its own unique characteristic curves and operating parameters. The factors affecting the choice of circuit configuration include: the input and output impedance that is needed; the voltage, current, and power gains that are wanted; and the maintenance of the transistor within its heat-dissipation limits. Also, the effect of added components such as capacitors and inductors must be considered, since these components will frequently limit the frequency range of the amplifier. These added components are often needed to isolate the signal from direct-current voltages and to bypass the signal around resistances. The bibliography at the end of this book lists several books that go into more detailed descriptions of transistor operation and transistor circuit design.

SILICON-CONTROLLED RECTIFIER

The silicon-controlled rectifier (SCR) is a four-layer device (NPNP) with three leads: a cathode lead, an anode lead, and a gate lead (see Figure 1.20).

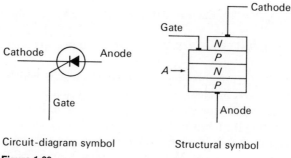

Circuit-diagram symbol   Structural symbol

Figure 1.20
High-power silicon-controlled rectifier (SCR).

The SCR will operate as a rectifier only when the reverse bias across junction A (when so biased that the anode is positive with respect to the cathode) can be broken down. At a particular forward voltage known as the breakover voltage, the SCR will begin to conduct between anode and cathode like a rectifier. By using the gate lead, however, it is possible to make the SCR conduct without having to raise its voltage to the breakover point. By applying a very small current between the gate lead and the cathode (by connecting the gate lead to the anode) the SCR will be triggered "on"; that is, the gate signal enables current from the anode to cross the reverse-biased junction A. Once the SCR begins to conduct, it will continue to do so even if the gate is disconnected.

The circuit of Figure 1.21 shows how the SCR may be wired to act as a sensitive switch. If a wire is momentarily touched between point B and the SCR gate G, current will flow from the gate to the cathode, the SCR will be triggered into conduction, and the lamp will light. The SCR will continue to conduct after the gate lead is removed. To extinguish the lamp, it is necessary either to remove the battery from the circuit or to momentarily connect point B to point D. The latter connection short-circuits the SCR and brings it under gate control again.

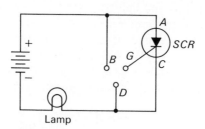

Figure 1.21
SCR circuit used as a
sensitive switch.

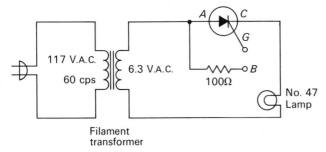

Figure 1.22
SCR circuit used as an electronic relay.

The SCR can also be used with alternating currents. Figure 1.22 shows a SCR used as a sensitive relay. When the gate lead *G* is touched to point *B*, the SCR will be triggered on and will conduct like a diode rectifier; that is, only when the positive-going half-cycles are applied to the rectifier's anode *A*. As long as the gate is connected to point *B*, the SCR will continue to conduct the positive half-cycles. When the gate lead is separated from point *B*, the SCR will complete the conduction of the last positive half-cycle and then stop conducting. Thus the lamp or other load can be controlled exclusively by the gate.

The SCR is a high-speed device capable of switching from a nonconducting to a conducting state within a few microseconds. This fast switching ability, and its ability to handle high power, make the SCR particularly suited for use in power inverters, power controls, and other circuits that may operate at frequencies that are too fast for electromechanical relays.

OTHER SEMICONDUCTOR DEVICES

There are many types of semiconductor devices designed for special purposes. Some of these devices are the light-activated SCR, the unijunction transistor, the snap diode, tunnel diodes, the dynaquad, and the binister. Each of these is made from special combinations of semiconductor materials that result in the operating characteristics desired. Descriptions of the operation of these devices may be obtained from the manufacturer or from textbooks, but are beyond the scope of this book.

INTEGRATED CIRCUITS

Integrated circuitry is one of the latest developments in the field of electronics. Specialized manufacturing techniques allow entire circuits, consisting of the equivalent of several transistors, resistors, capacitors, and diodes, to be created in containers the size of a single conventional transistor. Although a variety of techniques are used to create each integrated circuit (IC), the typical manufacturing sequence involves the etching of individual components out of a tiny, solid silicon chip. This technique results in a rugged component known as a monolithic IC. Other techniques are the evaporation of microscopic material onto a ceramic base, known as thin-film ICs, and the printing of overlapping layers of ceramic and metallic materials, known as thick-film ICs.

Integrated circuits are now used extensively in computer circuitry for gating and counting functions. In addition, entire operational, audio, r-f, and video amplifiers are available in single integrated circuits. Although a detailed description of these circuits is beyond the scope of this book, several integrated-circuit projects are included.

2

Understanding Circuit Diagrams

There are several basic types of diagram used to describe the layout, construction, and operation of electrical devices. Figure 2.1 shows four of these diagrams; each one describes some aspect of the same device, which is a versatile electronic amplifier-relay that can be used to sense a variety of activities.

SCHEMATIC DIAGRAM

A schematic diagram is a map of a specific circuit in a standardized form (see Figure 2.1, a). Symbols represent all the active circuit components used in the construction of the device, and the necessary interconnections of these components. In addition, a key to the components (i.e., R, Q, S, etc.) is provided. These symbols refer to specific components, whose description may be found in the parts list, which will usually accompany the circuit diagrams in this book and will list the specific values of all the components required, to enable the builder to order the correct components directly from a parts supplier. The parts list for the circuit in Figure 2.1 is as follows.

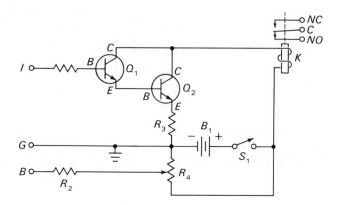

(a) Schematic diagram

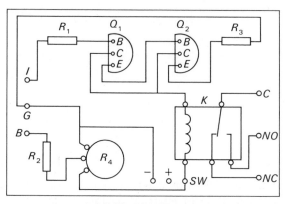

(b) Pictorial diagram

Figure 2.1
Basic circuit diagram showing: (a) detailed schematic diagram;
(b) pictorial diagram showing parts placement;
(c) exploded-view diagram showing assembly details;
(d) block diagram showing interconnection of
device to be used as a drinkometer.

R_1. 1,000-ohm, $\frac{1}{2}$-watt, 10%.

R_2. 10,000-ohm, $\frac{1}{2}$-watt, 10%.

R_3. 47-ohm, $\frac{1}{2}$-watt, 10%.

R_4. 1-megohm, 1-watt potentiometer.

Q_1, Q_2. 2N3708.

K. 1,000-ohm sensitive relay (Sigma 4F1000 or equivalent).

S_1. SPST switch.

B. 9- to 12-volt battery.

Miscellaneous: Chassis, wire, hardware, terminals.

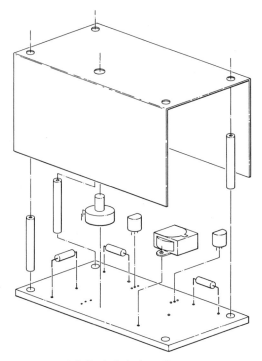

(c) Exploded-view diagram

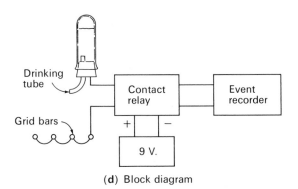

Drinking tube

Grid bars

Contact relay

Event recorder

+ −

9 V.

(d) Block diagram

The schematic diagram does not necessarily show how the components should be placed, although often the circuit would perform adequately if the schematic diagram were followed for the physical placement of the parts. Other circuits would not perform adequately, since some components and wires have electric fields that could interact with nearby components. Also, some components generate heat, which could prove destructive to other components. However, in the circuits described in this book, the physical location of the components in the completed circuit will rarely be critical. When it is, component placement will be prescribed in the construction notes that accompany the circuit diagram.

PICTORIAL DIAGRAM

The pictorial diagram shows how the components are arranged and interconnected in accordance with the schematic diagram (see Figure 2.1, b). The pictorial diagram is either a two- or three-dimensional view of the actual circuit and is a valuable aid in construction. It is often used to suggest a layout for printed-circuit construction, which will be discussed in Chapter 3. The pictorial diagram will relate the actual size and location of the components to the symbols on the schematic diagram. In this book the pictorial diagram will show a preferred or necessary way to lay out components.

EXPLODED-VIEW DIAGRAM

The exploded-view diagram is a three-dimensional view of how the circuit and chassis parts are assembled into the completed device (see Figure 2.1, c). Throughout this book the exploded-view diagram will be used to illustrate construction techniques that are not readily apparent from a study of the schematic or pictorial diagrams, in particular, how to mount certain solid-state devices, such as a control rectifier, on a heat sink in order to assure proper heat dissipation but without causing short circuits. It will also be used to illustrate novel methods of assembling components, such as a light cell and lamp, into unique circuit elements for special functions.

BLOCK DIAGRAM

The block diagram shows how the device may be used in specific experimental setups (see Figure 2.1, d). It is also useful for describing functional areas within a circuit, such as amplifier stages, oscillator stages, and output stages. It will be used for both purposes throughout this book.

CIRCUIT COMPONENTS AND SYMBOLS

Before we discuss the construction and functioning of a circuit, we must consider some additional information in order to comprehend more fully the schematic diagram and what it represents. Returning to the schematic diagram (Figure 2.1, a), notice that the inputs to the device (i.e., those parts labeled *I*, *G*, and *B*) are located on the left side of the diagram, and the outputs of the device (i.e., *C*, *NC*, and *NO*), on the right side. This convention is followed by designers in creating a circuit diagram and provides the first clue as to how a circuit functions. In addition, standard symbols are used for component parts such as resistors, capacitors, relays, transistors, and diodes. Appendix A lists many of the components used throughout this book and shows the corresponding schematic symbols. The numerical values of some components as given in a parts list are not obvious, and so are color-coded on them according to the codes in Appendix B. Examples include the colored bands around resistors and the colored dots on the face of molded capacitors. This color coding enables the builder to identify a component part and easily relate it to the schematic-diagram symbol and the parts list.

WIRING CONVENTIONS

Figure 2.2 illustrates how wires that cross each other and that intersect each other may be represented on a schematic diagram. In most schematic diagrams one crossing-wire convention and one intersecting-wire convention are maintained throughout the diagram for clarity. The two inter-

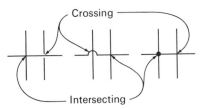

Figure 2.2
Crossing and intersecting
wire conventions.

secting-wire conventions that appear to show wires being routed spatially (i.e., above and below each other) are intended not to represent physical wire placement in a completed circuit, but only to facilitate an understanding of the circuit.

GROUND AND CHASSIS SYMBOLS

In order to reduce the number of wires presented on a schematic diagram and to illustrate the use of the chassis as a conducting element, the conventions in Figure 2.3 are used. Figure 2.3, a, indicates a common return point for all components and wires that terminate in this symbol. The term "ground" is actually a carryover from the early days of radio, when the receiver was literally connected to the earth (ground). A conven-

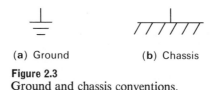

(a) Ground   (b) Chassis

Figure 2.3
Ground and chassis conventions.

tion from the early days of electronics referred the negative circuit voltage to ground. This convention is no longer applicable in many circuits. To avoid any misunderstanding, the ground connection is now usually established by listing the polarity of the source voltage on the schematic diagram. Other circuits refer to a positive, negative, and common voltage, and it is standard practice to label the common point as ground. Figure 2.3, b, indicates the common use of the chassis as a return point for wires and components.

In some schematic diagrams a bus line is indicated as the common return line for various components and wires. All circuits connected to the bus are returned to ground instead of being connected directly to the chassis. This grounding is often required to avoid hum-producing ground loops, which could interfere with the conduction of a signal through a circuit. A bus line is usually indicated by an extra heavy line on the schematic diagram.

Two examples of ground loops are shown in Figure 2.4. In Figure 2.4, a, the chassis may actually function as an added resistance, $R_{chassis}$, in the oscillator circuit, thereby affecting the tuning of the circuit. The chassis may also transmit undesirable ground currents generated in other parts of the circuit; these currents might easily be coupled to the oscillator circuit,

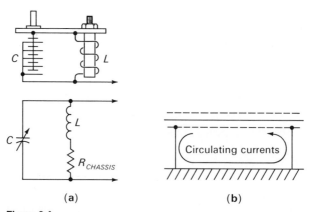

(a)   (b)

Figure 2.4
Examples of ground loops: (a) Improper oscillator-circuit grounding; (b) Improper coaxial-cable grounding.

and distort it. In Figure 2.4, b, a coaxial cable is shown with both ends of its shielded cable connected to the chassis. If currents are circulating through the chassis because other circuit elements (such as the oscillator circuit of Figure 2.4, a) are connected to it, these currents could circulate in a loop through the shielded wire of the cable and generate undesirable signals in the signal wire, resulting in hum or distortion of the signal. To prevent this, coaxial cables should normally be grounded to the chassis at only one place.

CONNECTIONS TO VOLTAGE SOURCES

Figure 2.5 shows three of the many conventions that are used to refer one part of a schematic diagram to other parts of the same diagram. These three are usually employed to show how various parts of the circuit are connected to sources of voltage. Figure 2.5, a, shows three sources of

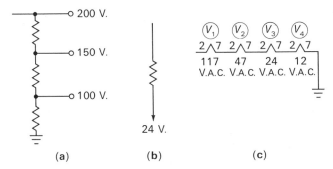

Figure 2.5
Voltage-source conventions.

voltage terminated by large black dots. Corresponding dots and voltages can be found elsewhere on the schematic diagram and refer to a connection between like voltages. Normally a wire could be drawn between like voltage sources, but this may clutter the schematic diagram, since there may be many points on the schematic to which a specific source of voltage must be connected. An arrowhead may be similarly used to illustrate where a particular voltage must be connected (see Figure 2.5, b). The connection that receives the voltage will be labeled with that voltage, and may consist of either an arrowhead or an arrowtail.

Figure 2.5, c, shows the reader how to wire a series of vacuum-tube filaments to a 117-volt A.C. source. V_1, V_2, V_3, and V_4 are vacuum tubes shown elsewhere on the schematic diagram; the triangles labeled 2 and 7 refer to the pins on the vacuum tubes that receive the wires.

CABLE WIRING SYMBOLS

Figure 2.6 shows three ways to represent bundles of wires or shielded wires. Figure 2.6, a, represents a shielded or braided cable with one conductor. The shield usually runs the entire length of the wire and, as illustrated, is grounded. Figure 2.6, b, is another way to represent such a cable. Multi-

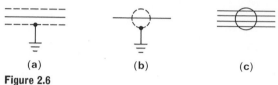

(a)   (b)   (c)

Figure 2.6
Cable-wiring conventions.

conductor shielded cables would be shown with the shield surrounding the entire group of conductors. Figure 2.6, c, shows a bundle of wires tied together by an outer covering that is not shielded.

FROM SCHEMATIC TO COMPLETED CIRCUIT

Although it is not necessary to understand in detail how a circuit works before one begins to construct it, a general idea of how it works could be of great assistance in the construction. In this book all circuits will be briefly described after their applications and specifications are detailed. Likewise, a pictorial diagram of the circuit may help in construction, but is not usually necessary. Where detailed pictorial diagrams are required for the construction, they will be given in the construction notes.

All that is needed to construct the electronic relay of Figure 2.1 is the circuit diagram and the parts list. The following steps will illustrate one method of preparing for and actually constructing the circuit of Figure 2.1. Detailed construction methods will be discussed in Chapter 3.

1. Procure all the parts needed for constructing the circuit. Notice that several of the listed parts are included under "miscellaneous": wire, input and output terminals, case, grommets, and so on. The details of these parts are not specified, because they are interdependent. For example, some pieces of equipment might have snap terminals, which could dictate the use of similar terminals in the device under construction, which in turn may require that a plastic case be used to house the circuitry, since snap terminals are difficult to insulate from a metal case. Similarly, it may be desirable to locate control R_4 at some distance from the rest of the circuit; this would necessitate long runs of wire or possibly a three-conductor cable. In general, however, 20-gauge hookup wire will be adequate to completely wire the circuit (as well as most of the circuits described in

this book). Another item needed before construction can begin is a suitable foundation on which the individual components may be mounted and interconnected. Although the parts list specifies a printed circuit board, other means of connecting components may be employed. Chapter 3 lists several methods of interwiring components, details the advantages of each method, and describes the actual process of soldering and interconnection.

2. Establish the layout of those parts that will be attached to the case of the device, i.e., terminals, switches, controls, etc. Often the layout will be dictated by the use to which the device will be put. Careful consideration should be given to assure an easily accessible final layout.

3. Study the schematic diagram with the idea of economizing on wiring, which saves time in the actual construction, results in a compact, neat circuit, and is especially helpful when modifications or repairs are required. An example of wiring economy is shown in Figure 2.7. It can be seen in Figure 2.7, b, that no hookup wire is used to interconnect components. The leads from the components themselves are wired directly to each other to provide the interconnections specified in the portion of the schematic diagram shown in Figure 2.7, a.

4. Constantly check the progress of the construction against the circuit diagram. It is advisable not to solder any components together until all the interconnections to each component are attached. Notice in Figure 2.7, b, for example, that the wiring to the connector at point B is not soldered, since other wires will be attached to it.

5. Observe good construction practices as discussed in Chapter 3.

6. When the circuit is fully assembled, check and then recheck the circuit against the schematic diagram, other construction diagrams, construction notes, and caution notes before applying power to the circuit.

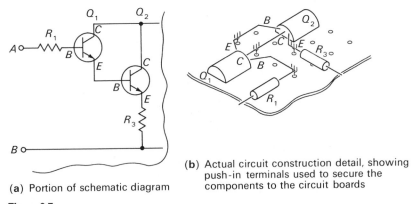

(a) Portion of schematic diagram

(b) Actual circuit construction detail, showing push-in terminals used to secure the components to the circuit boards

Figure 2.7
Method of wiring using component leads instead of hookup wires.

3

Construction Techniques

With the development of transistor technology have come simple, stand-ardized techniques for building and reproducing electronic circuits. These techniques evolved from the need to create small, reliable, and easily reproducible circuits consistent with the inherent miniaturization of solid-state devices. Perhaps the most important of these construction techniques has been the printed or etched circuit. The printed circuit is a way to both provide a foundation for component parts and eliminate the need to inter-wire them. The printed circuit contains one or more copper conductor layers that are selectively etched into a circuit configuration. As the name implies, printed circuits can be easily reproduced with uniformity.

With the development of easily reproducible circuitry have come tech-niques that enable the experimenter or circuit designer to create prototype circuits that will approximate the packaging and performance of manufac-tured devices. These prototype circuits are known as breadboard circuitry; in the early days of electronics, experimental circuits were constructed on wooden bases that were said to resemble kitchen breadboards. These bread-board circuits enabled the experimenter to modify very easily the circuit being constructed, since the components were not fastened permanently to the breadboard and could easily be replaced. With the introduction of solid-state technology, the old-fashioned breadboard circuitry techniques evolved into special methods of construction. Many of these methods still permit simple changing and rearranging of components, although the physical size of the breadboard foundation has been considerably reduced.

Construction techniques familiar to those who have assembled vacuum-tube circuits are also used with solid-state devices. One of these techniques is point-to-point wiring. By this method component parts are attached to each other using terminal strips, solder lugs, wires, and a central chassis for the entire circuit. Another technique uses a circuit board, or subchassis, which can hold interwired components and which in turn is fastened to a main chassis. Several circuit boards may be attached to a common chassis, with the circuit boards interwired into a total circuit configuration. All these construction techniques will be discussed in detail in this chapter, along with techniques of soldering and of handling semiconductors, and safety precautions to be observed when working with electrical devices.

PRINTED CIRCUITS

Printed circuitry is used extensively by manufacturers for mass-producing electronic circuits. It also provides a valuable method of construction for the researcher or experimenter who wishes to create a single reliable circuit, especially in constructing circuits where rugged miniaturization is desirable, such as a remote telemetry transmitter to be implanted in an animal. In addition, printed circuits are valuable when improper placement of component parts could result in hum and overheating, as in the construction of preamplifier circuits.

A printed circuit consists basically of a copper-clad sheet of plastic. The copper is bonded to the plastic and then etched according to the schematic diagram into conducting paths for electricity. Holes are drilled through the copper and the plastic to make it possible to mount the components on the circuit board and to solder them to the copper. Figure 3.1, a, shows a completed, etched circuit board with the holes drilled. Figure 3.1, b, shows a completed circuit, with the component parts soldered to the printed circuit.

It is also possible to create printed circuits that have copper on both sides of the plastic sheet. This method permits an even greater packaging density than the use of a single copper sheet, since far more conducting surface can be fitted into a given area.

PRINTED-CIRCUIT CONSTRUCTION

The actual construction of a printed-circuit board involves drawing the desired circuit and transferring this drawing to the copper-clad board.

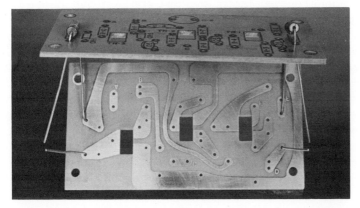

(a)

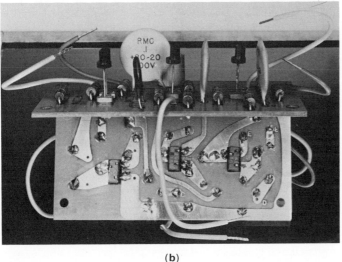

(b)

Figure 3.1
(a) Etched, printed circuit board, labeled to show desired parts placement; (b) component parts affixed to the circuit board. (These were photographed on a mirror to show the bottom views also.)

Many techniques are used to transfer the drawing and then etch the board to conform to it. In addition, there are certain criteria for determining how close adjacent lines should be placed to each other, and how wide the etched lines should be. These criteria will not be discussed in this book, since, where printed-circuit construction is recommended, a layout of the pattern will be included in the construction notes accompanying the circuit description.

Two simple methods for transferring a printed-circuit drawing and then etching the circuit board are as follows. One method, called the tape-resist method, involves drawing the circuit pattern with either a special tape or a specially prepared resist solution, both of which are impervious to an etchant solution. The entire board can then be immersed in the etchant solution, which dissolves the copper from all parts of the circuit board that are not protected by the tape or resist solution.

There are several manufactured printed-circuit kits that contain all the materials required to completely etch a circuit board, using a combination of resist tapes and resist solutions that are painted directly onto the copper laminate. These kits enable the experimenter to create a finished printed-circuit board ready to have holes drilled for the mounting of components. This method of construction is particularly useful when the required circuit is not to be duplicated and where rapid, efficient, and inexpensive construction is required for a particular circuit.

The second method, called photoetching, uses a photosensitive copper-clad laminate. The copper-clad sheet is covered with the photosensitive material and is exposed to light through either a hand-cut stencil negative, which is cut with the desired circuit pattern, or a specially prepared photo-negative containing the desired circuit pattern. It is considered a reverse photoetching technique because the area of the copper that is exposed to light through the stencil or film is made insensitive to the subsequent etching, whereas those areas that are not exposed can be etched. Following the exposure, the photosensitive laminate is fixed in a solution that effectively dissolves the sensitizing film from all areas of the laminate that were not exposed to light, leaving the desired pattern insensitive to the subsequent etching. The etching solution removes the unexposed copper, leaving the completed printed circuit ready for drilling of component holes.

Several complete photoetching kits are manufactured. These kits contain all the materials and instructions needed to enable the experimenter to cut his own stencil and create several printed circuits from the original stencil. In addition, the manufacturers of these kits will usually provide, for a nominal cost, a photonegative from original art work, if the experimenter finds it impossible or undesirable to cut his own stencil. With these kits, the experimenter can complete small production runs of printed circuits. I used this method to build twenty contact relays that were subsequently used as drinkometers in a series of experiments. The photoetching produced uniform circuits with a high degree of reliability.

BREADBOARD CIRCUIT KITS

A variety of breadboard kits are available to the experimenter. Some can be used for creating temporary circuits with removable components, others for creating permanent circuits with soldered components. The latter circuits are useful and reliable, and perform like manufactured circuits on printed-circuit boards. The temporary kits are extremely helpful in creating test circuits where components may be easily substituted and various component layouts evaluated without having to solder and desolder whenever a change is desired. With some of these kits, the circuit can be assembled and tested in a temporary layout and then soldered for permanent installation.

Figure 3.2
Vectorbord circuit, showing components temporarily
mounted on spring connectors.

A typical temporary circuit made from a popular breadboard circuit kit manufactured by the Vector Co. is shown in Figure 3.2. The circuit is constructed using push-in spring connectors that hold the components without soldering. The connectors in turn are inserted into a prepunched plastic sheet known as Vectorbord, which acts as the foundation for the entire circuit. In addition to the spring connectors shown in Figure 3.2, a variety of other connectors are available, including tube and transistor sockets that permit temporary wiring to other circuit components.

A typical permanent circuit is shown in Figure 3.3. The connectors for this circuit consist of push-in terminals with serrated edges designed to grip component wire leads. When inserted into Vectorbord, these connectors enable a circuit to be constructed and tested in a temporary layout. The components may then be soldered directly to the terminals for final

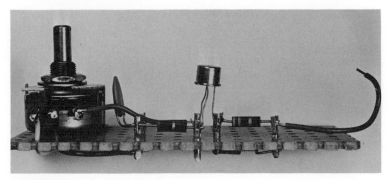

Figure 3.3
Permanent Vectorbord circuit.

installation. This breadboard circuit kit is also manufactured by Vector and consists of a variety of connectors, including spring connectors, connector insertion tools, tube and transistor sockets, bus wires, brackets for switches and controls, and several sheets of Vectorbord.

Figure 3.4 shows a circuit constructed on a special copper-laminated plastic surface known as Veroboard. The Veroboard, a type of universal printed circuit, is a prepunched board with copper strips running its length. Components are inserted into the plastic side of the board and then soldered directly to the copper strips, which form the conducting surfaces. A special tool is provided with the Veroboard for cutting the copper strips

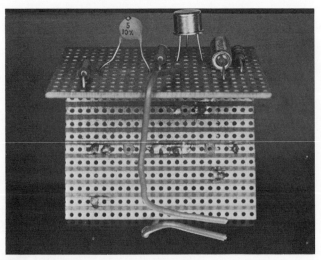

Figure 3.4
Verobord circuit (photographed on a mirror to show bottom view also).

where required. This method of construction is very similar to standard printed-circuit construction and is extremely valuable for creating a reliable finished circuit.

Other breadboard kits consist of perforated plastic boards with pre-mounted solder-filled eyelets. Components are inserted into the eyelets by melting the solder in the eyelet and letting the component lead penetrate through the hole. When the solder cools, the component is held securely in the eyelet. Still other kits consist of specially designed solderless connectors that allow simple and efficient component replacement.

There are over twenty different breadboard kit manufacturers; descriptions of the various breadboard kits available can be found in industrial parts catalogs. Each of the kits offers some advantages over the others, depending on the applications. For example, kits using spring terminals eliminate the chance that sensitive components will be damaged by the heat of repeated solderings, but may prove impractical for building circuits that operate at high frequencies, since excess wire may act as an antenna or inductor. It may prove advisable to obtain several different kits for a variety of projects.

POINT-TO-POINT WIRING

One of the most common methods of circuit wiring is point-to-point wiring, with the chassis serving as a ground and with components attached to the chassis rather than to a circuit board. The components are wired to each other by means of terminal strips that act as a central collection point for wires that are to be connected. An example of point-to-point wiring is shown in Figure 3.5. This method of wiring is generally used for vacuum-tube and relay power-supply circuits, where the components (tubes, sockets, electrolytic capacitors, etc.) are relatively large and there is enough space to wire components together. Although this method of wiring may also be used for low-voltage, low-power solid-state circuits, it is often impractical to construct such circuits. The relatively small components either would be crowded together or would need long wires attached to them. The result would be a complicated layout, difficult to service in case of malfunction. For this reason, solid-state circuitry is usually constructed on circuit boards, which are then mounted on the chassis and interwired either with other circuit boards or with input-output connectors, power connectors, or other components, all in a point-to-point manner, using terminal strips and lugs.

CIRCUIT-BOARD WIRING

Printed circuits and breadboard circuits are only two types of circuit-board wiring; there are others, such as the use of Vectorbord foundations,

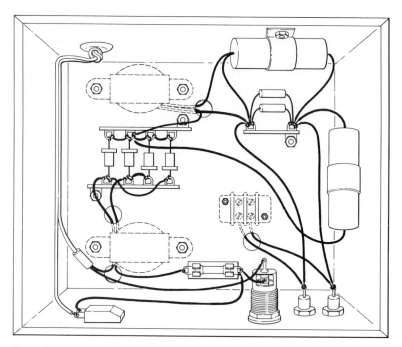

Figure 3.5
Point-to-point wiring.

which can accommodate component leads and wire. The Vectorbord foundations used in circuit-board wiring differ from those used in breadboard wiring. As noted in the section on breadboard circuit kits, Vectorbord can serve as a foundation for spring clips, which enable components to be held without soldering. In circuit-board wiring, the Vectorbord acts as a foundation upon which components are mounted as illustrated in Figure 3.6, a. The component leads, which pass through the perforations, are routed together in accordance with the circuit diagram (Figure 3.6, b) and soldered together as illustrated in Figure 3.6, c. Where relatively small components are employed, the component leads themselves can usually be used to connect the components. This method of construction results in a neat, compact layout, simple to troubleshoot in case of malfunction.

A circuit board for circuits operating with low voltages may be constructed with ordinary cardboard as a foundation. An example of a circuit created in this manner is shown in Figure 3.7. A printed circuit-board pattern (Figure 3.7, a) is first drawn on the cardboard and holes are then punched in it. The components are inserted into the holes and the component leads carefully soldered together (Figure 3.7, b). When completed, the entire circuit may be sprayed with a plastic film to make the cardboard surface waterproof and to secure the components.

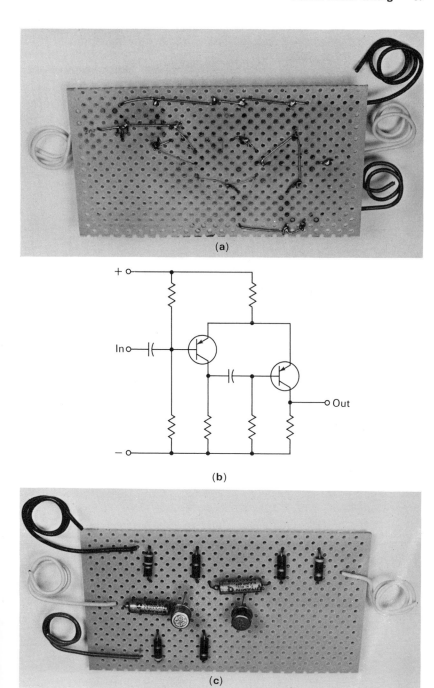

Figure 3.6
Circuit-board wiring following circuit-diagram layout.

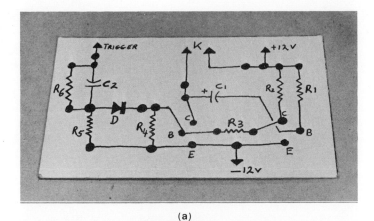

(a)

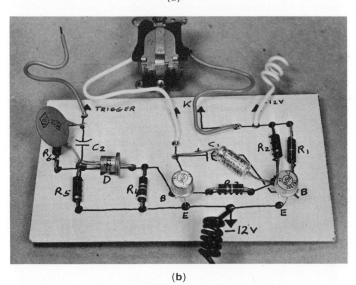

(b)

Figure 3.7
Cardboard circuit-board wiring following circuit-diagram layout.

TECHNIQUES OF SOLDERING

Proper soldering technique may be the factor that determines whether an electronic circuit operates properly or not. If the builder has had no experience soldering electrical connections, he should take the time to gain some experience by practicing on scrap materials.

Selection of a Soldering Tool

There are a variety of soldering irons and soldering guns used to make secure electrical connections. Figure 3.8 shows three of these soldering tools, each designed for a particular application. The soldering gun (Figure 3.8, left) is a fast-heating tool generally used for soldering heavy electrical connections, but that can also be used for many other tasks, such as cutting plastic, soldering copper pipe, and repairing broken metal. These guns have wattage ratings between 100 and 400 watts and are relatively heavy, rugged tools. They are extremely difficult to use when soldering small connections in tight locations. In addition, their large heat ratings make them particularly dangerous to use when soldering transistors, which may be damaged by excessive heat.

For general electronic-circuit construction, a lightweight soldering iron, such as is shown in Figure 3.8, center, is advisable. These irons have wattage ratings between 20 and 50 watts and are easy to handle when soldering small connections in difficult locations. The tip shown in Figure 3.8, b, is used for most soldering; but interchangeable, differently shaped tips are available for other applications, such as printed-circuit soldering, where there is a very small working area.

The pencil iron in Figure 3.8, right, is particularly useful for miniature work, such as integrated-circuit soldering and printed-circuit soldering. Pencil irons are low-wattage units, usually between 6 and 25 watts, and

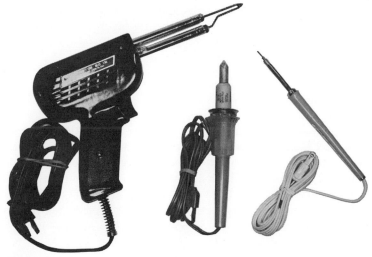

Figure 3.8
Soldering tools.

are the most appropriate device for soldering leads on transistors and other components that are sensitive to heat or are extremely small. These irons are inappropriate for soldering wires to switches or solder lugs, because the area to be soldered is usually much larger than the area of the pencil tip.

Soldering irons and guns should be tinned before being used, i.e., should be covered with a layer of solder. First heat the tip and then touch it with solder, so that the entire area is covered with the solder. Shake the iron or gun to slip off the excess solder. Repeat this process until the tip is uniformly covered, insuring that the tool will work effectively. In addition, it is advisable to clean the soldering tip, when it appears to be corroded or pitted, with a damp cloth or sponge, a file, or steel wool, before tinning it. Most soldering tips are made of copper, although there are special tips that are not supposed to corrode and that should not be filed.

Selection of Solder

The correct solder for all electric-circuit soldering is known as resin-core solder. Resin-core solder is a metal alloy (usually tin and lead) impregnated with a resin that acts as a flux, enabling the solder to "wet" or adhere to the surface of the joint being soldered. Solder is available in various sizes and alloy compositions. In general, a good grade of 18-gauge resin-core solder composed of 60% tin and 40% lead is ideal for most electric circuits.

Solder with an acid core is also available, but should *never* be used for electric-circuit soldering. Although the acid is an excellent cleaner, it will corrode the surfaces of the joint being soldered and could ultimately destroy the soldered joint, resulting in malfunction.

Soldered Connections

A properly soldered connection is made by heating the joint with the iron until the joint melts the solder. The solder is melted onto the joint, not the iron. After sufficient solder has been added, the iron is kept on the joint until the solder appears to flow smoothly. The iron is then removed from the joint.

Figure 3.9 shows well-soldered and badly soldered connections. In Figure 3.9, left, the connection is poor either because the iron or the work was not hot enough, because not enough solder was used, or because the solder did not flow. Figure 3.9, right, shows a poor connection caused by too much solder being applied to the joint; the result is a pool of solder at the base of the connection. The connections of Figure 3.9, right

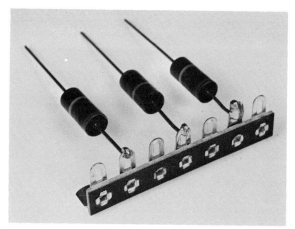

Figure 3.9
Soldered connections: **left,** too little solder;
center, good connection; **right,** too much solder.

and left, could result in cold-solder joints, that is, connections that appear solid but in reality are not connected at all. Cold-solder joints are difficult to discover when the circuit is completed and can be avoided by careful soldering.

Soldering Printed Circuits

A properly soldered connection to a printed circuit is made by first passing the component leads through the hole in the circuit board and bending them slightly (see Figure 3.10, a). This temporarily secures the component to the board. Next the iron tip is placed beside the component lead on the copper foil until the solder flows onto the component lead as shown in Figure 3.10, b. After enough solder has been added, the iron is kept in position until the solder fills the hole through which the component lead was passed. The iron is then removed from the board and when the connection has cooled, the excess of the leads is cut off (see Figure 3.10, c).

Soldering Semiconductor Devices

The leads on transistors and other semiconductor devices should be soldered as quickly as possible, to avoid damaging the device internally by excessive heat. Whenever possible, heat dissipators should be used between the semiconductor and the iron to conduct heat away from the component. Figure 3.11 shows three such heat dissipators. All three grip the component lead securely and free your hands to perform the soldering. The alligator

(a)

(b)

(c)

Figure 3.10
Method of soldering components to a printed circuit:
(a) inserting components onto board; (b) soldering
components; (c) completed soldered connection.

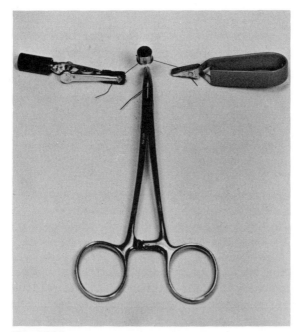

Figure 3.11
Heat dissipators.

clip in Figure 3.11, left, is modified by gluing tiny pieces of felt to the teeth
and dipping the tip in water before soldering.

SEMICONDUCTOR COMPONENT MOUNTING

Special heat dissipators, known as heat sinks, are often included in the
circuit to provide permanent cooling surfaces for functioning semiconductor devices that operate at higher than normal temperatures. These
dissipators usually consist of an aluminum surface to which as large a
surface of the component as possible is fastened. Both the size and the
shape of the heat sink depend on how much heat is to be dissipated from
the component. Commercially available heat sinks (Figure 3.12), consisting
of radiating fins that surround the component mounting area, provide for
optimum heat dissipation in a minimum of volume. Throughout this book
the heat sink will be specified either in the parts list or in the construction
notes.

MOUNTING COMPONENTS TO HEAT SINKS

Several factors must be considered when incorporating a heat sink into
a circuit. If the heat sink could short-circuit the component against the

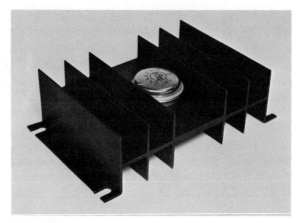

Figure 3.12
Heat sinks.

chassis or other circuit parts, it must be electrically insulated. The semi-conductor mounting hardware that is often included with components is designed both to protect components from short circuits and to assure

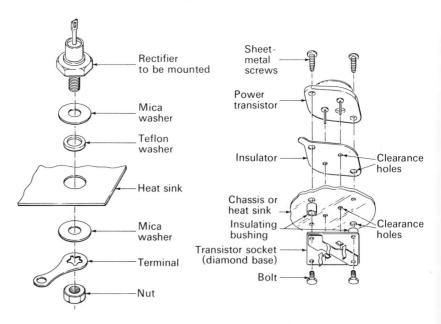

(a) Insulated rectifier
    or SCR mounting

(b) Typical power transistor
    mounting kit

Figure 3.13
Components mounted on heat sinks.

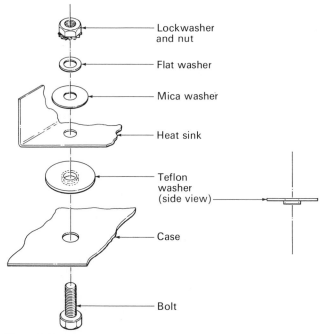

Lockwasher and nut

Flat washer

Mica washer

Heat sink

Teflon washer (side view)

Case

Bolt

Figure 3.14
Insulated heat-sink mounting.

adequate heat dissipation. If mounting hardware is not supplied, obtain insulated washers and other parts to adequately protect the component.

Two typical methods of mounting components to heat sinks are illustrated in Figure 3.13. Both methods provide for adequate heat dissipation and at the same time prevent short circuits from developing between the component and the metallic surface of the heat sink. With the rectifier mounting shown in Figure 3.13, a, the nut must be tightened onto the rectifier stud very carefully to prevent the stud from being stripped or stretched out of shape. Normally about 15 inch-pounds maximum torque applied to the nut will adequately secure the component. It is often necessary or simpler to insulate the heat sink from the chassis rather than from the component; Figure 3.14 illustrates a proper method for doing so.

LEAD-MOUNTED SEMICONDUCTORS

Lead-mounted semiconductor components may be secured by their leads. It is advisable when soldering leads to leave at least ⅛-inch clearance between the case of the component and the fastening point. It is often desirable to use transistor sockets to secure transistors to a circuit, since they eliminate the chance of damaging the transistor by soldering, and,

(a)

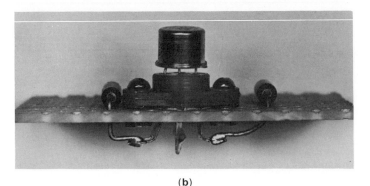

(b)

Figure 3.15
Transistor mounting; **(a)** soldered directly; **(b)** inserted in a socket.

should the transistor be accidentally damaged later, allow it to be easily replaced. Figure 3.15 shows a transistor soldered by its leads and another mounted in a socket.

SAFETY PRECAUTIONS

1. Be sure the power is off before working on a circuit.
2. When working on a circuit, make sure that electrolytic capacitors are discharged.
3. Be sure that circuits are properly insulated.
4. Solder carefully and avoid solder splashes.
5. Check to make sure that all solder joints are well-made.
6. Use fuses of proper ratings, and do not use slow-blow types unless they are specifically called for.
7. Use wire of the right size.
8. Provide for adequate cooling of components that heat up.
9. Keep hot components away from combustible materials.
10. Make sure someone is close by in case of an accident.

4

Troubleshooting Transistor Circuits

Transistors are physically rugged devices capable of withstanding high centrifugal and impact forces—they may be subjected to accelerations of 20,000 g without damage—but they are *extremely* sensitive to excess heat and to D.C. biasing voltages that are applied with the wrong polarity. When a transistor is subjected to more heat than it is rated for, its internal crystal-lattice structure is quickly destroyed; D.C. biasing voltages of the wrong polarity can cause excess heating. To understand how damaged transistors may be located, one must first have some basic understanding of transistor operation. (The reader may wish to review Chapter 1 before proceeding with the following discussion.)

Many types of testing apparatus for servicing transistor circuits are commercially available. This equipment includes several types of transistor checkers that can be used to determine whether the transistor is operative and how much gain or amplification it provides. Transistor checkers are particularly useful when one must match precisely the specifications of several transistors to be used in, for example, a differential amplifier. Some transistor checkers are specifically designed to test the functioning of transistors that operate at high frequencies or high switching speeds. It is useful to have, in addition to the test equipment especially made for transistor servicing, several standard pieces of test equipment, including a multimeter (VOM), a vacuum-tube voltmeter (VTVM), an oscilloscope, and a test oscillator. The VOM (volt-ohm-milliameter) combines three basic meters in one convenient package. When used as an ohmmeter, it can measure the resistance of circuits and components, and check the continuity of circuits. Its internal circuitry can be switched to function as an ammeter and as a voltmeter. It is by far the most commonly used piece of electrical test equipment.

Because meter movements themselves have electrical resistance, when a meter is attached to a circuit the resistance of the entire circuit will change because of the added resistance of the meter. Usually the change is insignificant, but it sometimes may be large enough to require the use of a meter with a high internal resistance. The vacuum-tube voltmeter (VTVM) is an electronic voltmeter with an internal resistance of about one million ohms. It has negligible loading effects on circuits and is commonly used to measure voltages in high-impedance electronic circuits where the less-sensitive VOM measurements would result in inaccurate readings.

The oscilloscope displays the shape of a voltage wave on the screen of a cathode-ray tube. Like the VTVM, it has a high internal resistance. It is useful for observing the distortion of wave forms in amplifiers, oscillators, and switching circuits, as well as for studying the relationships between simultaneous signals.

The test oscillator is a reference signal source. When used in amplifier servicing, it may be connected to the input of the amplifier, and the distortion of the oscillator signal observed as the signal passes through the various amplifier stages. The signal voltage may either be measured on a VTVM or be observed on the oscilloscope screen. The oscillator reference signal is usually either a square wave or a sinusoidal wave. The test oscillator signal can be varied in frequency and in voltage. Test oscillators commonly available provide audio-frequency and radio-frequency signals.

This chapter will primarily concern basic servicing techniques that can locate defective components either in a completed circuit or as separate parts. Although the techniques to be described may not be the most efficient or the most accurate, they can often isolate a defective component rapidly. The only test equipment needed for the tests outlined in this chapter will be the common multimeter (VOM) and the vacuum-tube or transistorized multimeter (VTVM). We will not discuss in detail servicing procedures that use other commercial test equipment. The bibliography lists several sources of information on such test equipment and procedures.

RECTIFIER TESTING

Current flow through a rectifier was illustrated in Chapter 1. In particular, it was shown how current could easily flow in the forward direction but not in the reverse direction. A common ohmmeter, such as that in a multimeter (VOM), may be used for a rough check of the ability of a rectifier to conduct electricity in the forward direction and to block conduction in the reverse direction. The ohmmeter is a device that delivers a small reference voltage to the circuit element being tested. It then measures the current flowing through the circuit element on a meter calibrated in ohms. If the leads of the ohmmeter are placed across an isolated rectifier, so that the positive ohmmeter lead is attached to the rectifier anode and the negative lead to the rectifier cathode, the ohmmeter will indicate a very low resistance because the rectifier is forward-biased and the only resistance being measured by the ohmmeter is the relatively small internal rectifier resistance. Conversely, with the ohmmeter leads reversed, the meter will indicate a relatively high resistance because the rectifier is now reverse-biased. Precise values for resistance vary with the type of rectifier, but the ratio of forward to reverse resistance is typically 500 to 1 for a medium-power rectifier.

This visual indication of rectifier conduction, known as front-to-back testing, is very useful for determining whether or not a rectifier has been internally destroyed and for determining which lead of the rectifier is the cathode and which is the anode. In making this latter check, the polarity of the ohmmeter leads must be determined either by an internal check of the ohmmeter battery polarity, by reference to the ohmmeter circuit diagram, or by using a voltmeter with known polarity to measure the voltage of the ohmmeter circuit. (Note that the lead polarity when a multimeter is switched to measure voltage or amperage may differ from the lead polarity when the multimeter is switched to measure resistance.)

TRANSISTOR TESTING

Although the internal physical structure of a transistor differs from that of the rectifier, it is convenient for testing purposes to consider the transistor to be two rectifiers with a common lead (see Figure 4.1); the PNP-type transistor can be considered to be two diodes with a common cathode, and the NPN type to be two diodes with a common anode. Front-to-back testing may then be performed on the transistor by the same principles as are used for testing the rectifier. However, before the ohmmeter test leads are applied to the transistor, the meter's current and voltage must be limited. Ohmmeters may operate at potentials ranging from 1.5 to 24 volts, and some transistors may be damaged by potentials greater than 6 volts.

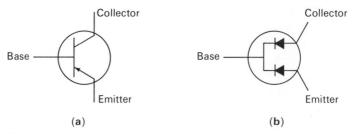

Figure 4.1
(a) Symbol for a PNP-type transistor; and (b) representative
diagram of a PNP-type transistor drawn as a two-diode device.

Fortunately, the typical VOM is constructed both with parallel resistors,
which limit the current, and with series resistors, which limit the applied
voltage, when the VOM is used to measure certain ranges of resistance. In
general, the highest resistance setting of the typical VOM will introduce the
most voltage to the circuit element being tested, and the lowest resistance
setting will introduce the most amperage. To avoid voltages and currents
that could be damaging when testing transistors, the ohmmeter should be
set to either the $R \times 10$ or the $R \times 100$ scale.

The following tests may be performed on low-, medium-, and high-power
transistors. The results of these tests, although not conclusive, can be
useful in rapidly locating defective transistors. All the tests are described
in terms of a PNP-type transistor. For an NPN type, it is only necessary to
reverse the meter leads.

Front-to-Back Testing, Base to Emitter

Ohmmeter readings may differ significantly for different types of
transistors, but the ratio of low to high resistance should always be about
1:500 (see Figure 4.2). For medium-power germanium transistors, typical

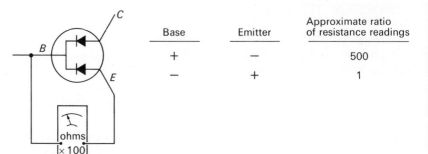

Base	Emitter	Approximate ratio of resistance readings
+	−	500
−	+	1

Figure 4.2
Front-to-back testing, base to emitter.

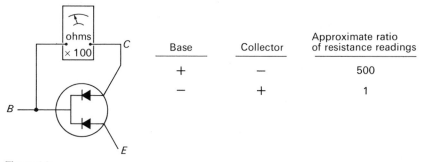

Figure 4.3
Front-to-back testing, base to collector.

base-to-emitter forward-biased resistance readings might be 100 ohms, the reverse-biased readings 50,000 ohms. (For front-to-back testing of base to collector, see Figure 4.3.)

Testing Cut-Off Ability

The PNP-type transistor may be cut off by applying the same positive voltage to the base and to the emitter (i.e., by shorting the emitter to the base). One notes the difference in resistance across the collector and emitter leads before and after the base is connected to the emitter (see Figure 4.4). Typical resistance readings for a medium-power germanium transistor might be 20,000 ohms at cut-off and 2,000 ohms with no base bias voltage. For high-power transistors, typical resistance readings might be 5,000 ohms at cut-off and 100 ohms with no base bias voltage.

Saturation Test

If the base bias voltage is negative with respect to the emitter, the PNP transistor will conduct electricity between the emitter and the collector.

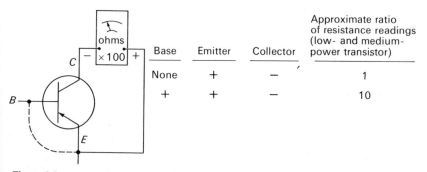

Figure 4.4
Transistor cut-off ability.

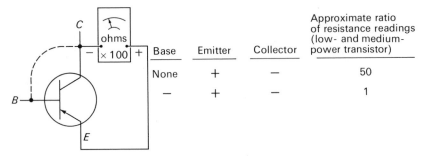

	Base	Emitter	Collector	Approximate ratio of resistance readings (low- and medium-power transistor)
	None	+	−	50
	−	+	−	1

Figure 4.5
Transistor saturation test.

Shorting the collector to the base causes a condition of maximum conduction, or saturation (see Figure 4.5). Typical medium-power transistors might exhibit resistance readings of 2,000 ohms with no base bias voltage and of 40 ohms at saturation. The ratio of the resistances is typically smaller for high-power transistors. High-power transistors might exhibit resistance readings of about 100 ohms with no base bias voltage and of 6 ohms at saturation.

To avoid destroying the transistor during this test, it is essential to limit the current flow to the base by keeping the VOM set to the $R \times 10$ or the $R \times 100$ scale, or by placing a bias resistor between the collector and base instead of shorting the collector to the base. Since most transistors operate with base bias resistors, placing a resistor in the circuit would more nearly simulate actual operating conditions. Any resistance up to about 200,000 ohms will produce meter readings.

The ratio of the change in the collector current to the change in the base current is known as beta, and is a measure of the amplification characteristics of a transistor. In the saturation test, the readings of the ohmmeter with the base lead open are compared with the readings taken with the base lead shorted (or connected through a resistor) to the collector. A transistor with a high beta will have less resistance with the base connected to the collector than a transistor with a low beta. This type of test is useful when many similiar transistors are being evaluated to determine which one has the highest beta.

IN-CIRCUIT TESTING

Testing procedures for transistors that are soldered into a circuit are similar to those for checking isolated transistors, but their results must be evaluated in terms of the circuit. A portion of a typical transistor amplifier circuit is shown in Figure 4.6. The PNP transistor may be evaluated as in the previously discussed tests, except that special consideration must be

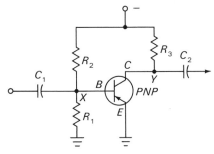

Figure 4.6
Detail of a typical transistor
amplifier stage.

given to the biasing resistors, R_1 and R_2, and to the collector resistor, R_3. The cut-off test may be performed with the negative VOM lead placed at point Y, and the positive VOM lead placed at ground. The power normally supplied to the circuit should be turned off, and the VOM internal battery used to provide operating power to the transistor. By shorting the base lead to the emitter lead, a resistance reading approaching the cut-off condition described under the cut-off test should be obtained. The results may not be identical, because the base and the collector are connected through resistors R_2 and R_3, but the base biasing resistor value is usually so high that the connection will not seriously interfere with the reading.

Saturation testing may be performed as outlined above, but consideration must be given to resistor R_1. Although R_1 may cause the resistance to be higher than usual when the base and collector leads are shorted, the reading should still be much lower than the nonshorted or normally biased resistance reading.

In-Circuit Voltage Testing

In-circuit tests may also be performed by observing voltage readings across circuit elements connected to the transistor stage being evaluated. Using normal operating power for the circuit, voltage readings may be taken and the results used to pinpoint damaged transistors, short circuits, and defective components. Voltage testing is more precise than ohmmeter testing for evaluating a circuit. Some circuit diagrams list test voltages that should be found in the actual circuit, and these may be compared with the actual voltages; any unusual differences in readings could lead rapidly to isolation of a defective component. An important aspect of voltage testing is that it is usually conducted with normal operating power applied to the circuit, since circuit trouble that will appear under normal operating conditions may not show up under the low voltages of ohmmeter tests.

In performing voltage tests on a transistorized circuit, it is important to use a test instrument with high input impedance. Impedance is the opposition to the flow of current in A.C. circuits, whereas resistance is the opposition to the flow of current in D.C. circuits. Impedance may be caused by circuit elements, such as resistors, inductors, or capacitors, and will be present in circuits that operate at relatively high on-off switching speeds. When measuring voltage in high-impedance circuits, one must use a measuring instrument that itself has high input impedance to the flow of current in the circuit being evaluated; otherwise the test instrument may function as a circuit element when placed in the circuit, the operation of the circuit could be severely altered, and one might obtain completely erroneous meter readings. The high-impedance voltmeters that are commonly available are vacuum-tube voltmeters (VTVM) or specially constructed transistorized voltmeters.

Testing for Broken Circuits

A simple one-transistor amplifier is shown in Figure 4.7. If the circuit should be broken at point A in the transistor base circuit, there will be no voltage reading across base bias resistor R_1. If it is broken at point B, there will be no voltage readings across R_1, R_2, or R_3, since all conduction through the transistor will have ceased. If it is broken at point C, there will be no voltage reading across R_3, but there will be readings across R_2 and R_1. However, the reading across R_2 will be less than normal, since there is no collector current and only a relatively small base current, and the reading across R_1 will be more than normal, because of the forward

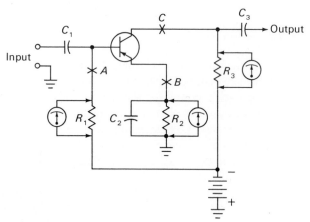

Figure 4.7
One-transistor amplifier.

bias from emitter to base with no collector current present to interact with the base circuit.

Testing for Shorted Transistors

If there is a short circuit between the emitter and base in Figure 4.7, there will be no voltage across R_3. The transistor will be effectively cut-off, reducing collector current to zero. (Very low voltage readings caused by leakage currents may be observed, but these can be ignored.) With a short circuit between the emitter and collector, there will be larger than normal voltage readings across R_2 and R_3, since the transistor is no longer controlling conduction, and since they are essentially connected in series across the battery (the effect of capacitor C_3 may be ignored). A short circuit between the collector and the base will cause saturation of the transistor, with higher than normal readings across R_3 and R_2, and a lower than normal reading across R_1.

Testing Transistor Conduction

The ability of a transistor to conduct a signal may be evaluated by measuring the voltage across R_3, first with the circuit undisturbed, then with the base lead shorted to the emitter lead; the short causes the transistor to cut off, and the voltage across R_3 should drop to almost zero. The ability of the transistor to control current is indicated by how close the reading comes to zero.

A second test is performed by altering the value of R_1 (possibly by placing a resistor equal in value to R_1 in parallel with it), which should result in a higher than normal reading across R_2. The reduction in the value of R_1 should cause a higher than normal base bias voltage, which in turn should cause the transistor to conduct more current through R_2.

Testing Components Other Than Transistors

Low-voltage electrolytic capacitors are used extensively in transistor circuits. They can be checked roughly, to see whether they conduct too much or too little current, with an ohmmeter set to limit the applied voltage. Their rated working voltages are often quite low, since they are designed for compatibility with the low voltages commonly used in transistor circuits. The voltage applied to these capacitors must not exceed their working voltage or they might be permanently damaged.

If an electrolytic capacitor is thought to be defective, one of its leads should be separated from the circuit and the ohmmeter leads placed across the capacitor. Be sure that the polarity of the ohmmeter lead matches that of the capacitor lead. (All electrolytic capacitors have one or both of their leads marked to indicate polarity.) The meter needle should deflect across the scale and then gradually return to the original (infinite resistance) reading, indicating that the capacitor has been charged by the ohmmeter battery. The smaller the capacity of the capacitor, the more quickly the needle should return to its original position. A broken capacitor will not produce a meter deflection. Although the above test may be performed on paper, ceramic, or mica capacitors, it becomes inconclusive when the capacitance value is too small.

In checking resistors that are connected to transistors in a circuit, one must consider the polarity of the ohmmeter leads when they are placed across the resistor. As discussed in the sections on transistor tests, the ohmmeter battery can power a transistor, causing conduction and consequent changes in the value of the resistor being checked.

A careful visual inspection of a transistor circuit should be considered an important step in any trouble-shooting procedure. The miniature components in transistor circuits may be easily damaged because of their close proximity to each other. Even though the individual parts are mechanically rugged, there is the chance that circuits will be shorted or broken by vibration or shock. One can often observe that components have been short-circuited. Another common circuit malfunction that results from vibration or shock is damage to the copper foil of a printed circuit. Breaks in the copper foil may be easily repaired by carefully melting solder across the broken portion, or by soldering a bridge wire between the two sections.

5

Power-Supply Circuits

Transistor circuits frequently require D.C. voltages for their operation. The required voltages come from power sources such as batteries, D.C. generators, or D.C. power supplies that operate on A.C. Although batteries and generators are common power sources, circuits that draw excessive currents will quickly discharge a battery or will require a large source of energy to drive a D.C. generator. Certain storage batteries, such as the automobile lead-acid battery or the nickel-cadmium type, may be used as power sources for circuits that draw relatively large currents, but if they are used continuously, a suitable charging circuit must be provided to recharge them. The charging circuit may consist of a D.C. generator or a power supply that rectifies A.C. into D.C.

By far the most common source of power used by the electronic experimenter is the D.C. power supply, which operates on 115-volt, 60-cycle, A.C. house power. The power-supply circuits to be described in this chapter all operate from house power. In addition to the D.C. power supplies, circuits are presented that change D.C. into A.C., and that change D.C. into D.C. of another voltage. The former circuit is known as an inverter, the latter as a converter. These circuits are used extensively in operations where battery power must be used. An inverter, for example, might be used in the field as a portable source of power for equipment that operates on A.C. The battery-operated converter might be used to operate a high-voltage shock source that must be electrically isolated from nearby sensitive measuring instruments.

The power-supply circuits presented differ from each other in the voltage and in the amperage they supply, and in the amount of filtering (elimination of A.C. variations, which are known as ripple) they perform in rectifying A.C. into D.C. Circuits designed to operate on D.C. may develop distortion when operated from power supplies with excessive ripple voltage.

The proper choice of power supply depends on the type of device to be powered and on the amount of current that will be drawn. Excessive load currents can reduce the output voltage of a power supply. The capacity of a power supply to maintain a constant, predetermined output voltage, even though the load or the input voltage varies, is known as regulation. An ideal power supply would maintain a perfectly constant voltage in spite of changes in the output current demanded by the load. The circuits presented in this chapter will illustrate several different methods of regulation, in the functional descriptions accompanying the circuit diagram. The specifications for each of the power supplies will list its maximum regulated output voltage.

POWER-SUPPLY CIRCUITS

Figure 5.1 shows the stages of a typical power-supply system. Not all of these stages are required in every application. The only essential ones are the A.C. source and the rectifying stage. The remaining stages contribute to making the supply more dependable; they keep ripple voltages small and insure good regulation.

The A.C. source provides the electrical energy to operate the load. It may consist of any source of alternating current; as stated earlier, the most common source is 115-volt, 60-cycle A.C.

The transformer changes the A.C. source voltage. It may step it either up or down. Since transistor circuits commonly operate with low voltages, step-down transformers are most common in transistor power supplies. Circuits using vacuum tubes usually require at least two operating voltages. A low voltage is used to heat the filaments of the vacuum tubes, and is

Figure 5.1
Elements of a typical power-supply system.

provided by a step-down transformer. The plate of a vacuum tube might require voltages of 300 or 400 volts D.C., derived from a step-up transformer, or sometimes from a voltage doubler circuit. Transformers used in vacuum-tube power-supply circuits may combine low-voltage and high-voltage secondary windings in a single package with a common primary winding.

Rectification changes the A.C. voltage to pulsating D.C. Various rectifier configurations are used to obtain pulsating D.C. with particular

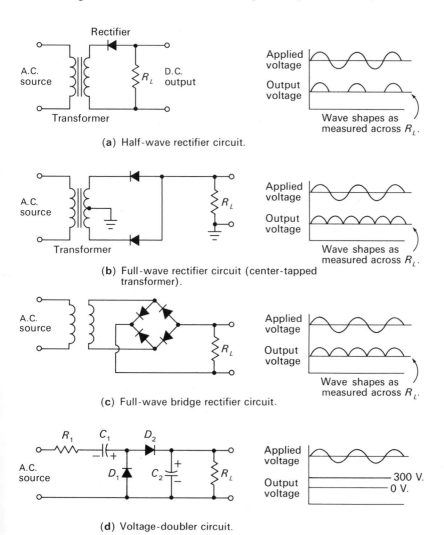

(a) Half-wave rectifier circuit.

(b) Full-wave rectifier circuit (center-tapped transformer).

(c) Full-wave bridge rectifier circuit.

(d) Voltage-doubler circuit.

Figure 5.2
Power-supply rectifier configurations.

characteristics. The most common of these configurations are shown in Figure 5.2, which also shows the wave shapes across a load resistor, R_L, that might be recorded by an oscilloscope.

The voltage-doubler circuit (Figure 5.2, d) is a special rectifier circuit designed to approximately double the A.C. input voltage. It is commonly used to obtain D.C. power without a transformer. A voltage-doubler circuit, along with a functional description of its operation, is presented later in this chapter.

The filter changes the pulsating D.C. from the rectifier to a smooth and constant D.C. Capacitors and inductors are most frequently used as filtering elements. A capacitor placed in parallel with the rectifier circuit offers very low impedance to the A.C. ripple frequency and very high impedance to the D.C. components. The ripple component is therefore bypassed to ground, while the D.C. voltage is applied to the load. An inductor in series with the load offers a high impedance to the A.C. ripple frequency and a low impedance to the D.C. components. Various combinations of capacitors and inductors will provide various combinations of voltage regulation and ripple filtering. A capacitor, or a capacitor followed by an inductor, does not regulate voltage as well as an inductor or an inductor followed by a capacitor, but considerations of cost become relevant here. Not only do inductors cost more than capacitors, but they can only operate with full-wave rectification, whereas the capacitor filter may be used with either half-wave or full-wave rectification. In general, inductor filters are used in circuits that must supply considerable power.

The type of rectification and filtering needed is determined by the characteristics of the circuit being powered. For general-purpose relay switching circuits, a half-wave rectifier and a single electrolytic capacitor filter will provide adequate power. For high-speed relay and solenoid switching circuits, a full-wave rectifier circuit coupled to an inductor or inductor-capacitor filter should be used. For relatively high-speed transistor switching circuits, a full-wave rectifier and a large-value electrolytic capacitor coupled to an electronic ripple filter should provide satisfactory operating power. Various combinations of capacitance and inductance are used to accomplish precise filtering and voltage regulation in special applications.

It is possible to filter ripple voltages very precisely by using an electronic filter in conjunction with a conventional filter. In a typical two-transistor filter, one transistor senses the ripple voltage and applies an amplified signal proportional to the ripple voltage to the base of a power transistor connected in series with the load. The power transistor conducts the D.C. to the load, and produces an A.C. voltage that is 180 degrees out of phase with the ripple voltage and cancels it out. A power-supply circuit employing electronic filtering can be found on p. 64.

Voltage-regulator circuits are designed to maintain a constant output voltage from a power source, even though the input voltage or the load varies. The zener diode (see Chapter 1) is commonly used in transistor power supplies to provide such a precisely regulated voltage. It may be operated in conjunction with one or two transistors to provide electronic voltage regulation. Circuits using electronic voltage regulation are described on pp. 62, 64, and 66.

The power-supply stages discussed above are used extensively in supplies that are constructed as integral parts of devices, as well as in separate supplies that are designed to operate several devices. Other techniques, more complex than these, can produce finer regulation and smoother output voltages. Separate reference amplifiers and comparison amplifiers, operated from separate power sources, are sometimes used to regulate a power supply. The circuitry may include methods of regulating and limiting amperage as well as voltage. Current regulators are sometimes used in timer circuits to assure accuracy by charging a timing capacitor with a constant reference current.

Commercially manufactured power supplies generally contain costly features that may not be required for a particular application. Such features as extremely fine ripple filtering, short-circuit-proof operation, and extremely fine voltage regulation may result in a power supply that is many times more costly than a supply constructed by the experimenter and tailored to his specific needs. The power-supply circuits presented in this chapter should serve the needs of most experimenters. Circuit diagrams and descriptions of more elaborate supplies may be obtained in textbooks devoted entirely to power supplies.

REGULATED POWER SUPPLY (9-VOLT)

Applications and Specifications

1. A well-regulated D.C. power supply, particularly suited for solid state timers, bridge circuits, and low-power transistor amplifiers.

2. Regulated output of 9 volts D.C. at 250 milliamperes. The accuracy of the voltage output depends on the tolerance of the zener-diode regulator ($\pm20\%$).

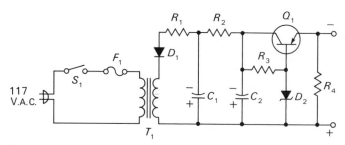

Figure 5.3
Circuit diagram of 9-volt regulated power supply.

The circuit diagram for this power supply is shown in Figure 5.3. The parts list is as follows.

R_1. 1-ohm, $\frac{1}{2}$-watt, 10%.
R_2. 2.4-ohm, 1-watt, 10%.
R_3. 150-ohm, $\frac{1}{2}$-watt, 10%.
R_4. 200-ohm, 1-watt, 10%.
C_1. 500-microfarad, 25-volt, electrolytic.
C_2. 500-microfarad, 25-volt, electrolytic.
D_1. Motorola rectifier, type HEP154.
D_2. Motorola zener diode, type HEP101.
Q_1. Motorola power transistor, type HEP200.
F_1. Fuse, $\frac{1}{4}$-ampere.
T_1. 12.6-volt transformer (Triad F-25x, or equivalent).
Miscellaneous: minibox, Vectorbord foundation, wire, red binding post, black binding post, fuse holder, push-in terminals, hardware.

Functional Description

A filament transformer supplies 12.6 volts to half-wave rectifier D_1. The current is filtered through a pi-type filter and applied through power transistor Q_1 and zener-diode regulator D_2 to the load. The zener diode and R_3 provide a reference voltage to the base of the power transistor and thereby effectively regulate the conduction of Q_1 throughout relatively wide variations of a load resistance. Changes in load current or input voltage to the regulator will cause a variation in the base-to-emitter voltage. If the voltage tends to rise, the transistor collector-to-emitter resistance will increase, thereby limiting the load voltage. Lowering the load voltage will tend to decrease the collector-to-emitter resistance, thereby increasing the load voltage. The net effect of this action results in the regulation of the output voltage to within a few millivolts of 9 volts.

Construction Notes

Use an aluminum or copper plate at least $3\frac{1}{2} \times 2 \times \frac{1}{16}$ inches as a heat sink for transistor Q_1. Take care to prevent short circuits between the heat sink and other circuits components (see Figure 5.4).

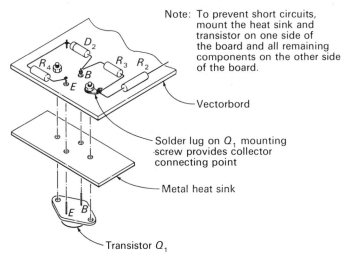

Note: To prevent short circuits, mount the heat sink and transistor on one side of the board and all remaining components on the other side of the board.

Vectorbord

Solder lug on Q_1 mounting screw provides collector connecting point

Metal heat sink

Transistor Q_1

Figure 5.4
Transistor mounting details.

VARIABLE LOW-VOLTAGE POWER SUPPLY (0.5 AMPERE)

Applications and Specifications

1. A general-purpose, laboratory-bench power supply that provides a continuous D.C. voltage of up to 20 volts, and is suitable for testing and evaluating circuits under construction and for general use where precise voltage regulation is not needed.

2. Load regulation: up to 12 volts with a 500-milliampere load, up to 20 volts with no load.

3. Ripple voltage: 0.5% between 2 and 9 volts, 1% between 9 and 12 volts.

4. Protected against overload damage by a power-line fuse and a circuit breaker.

The circuit diagram is given in Figure 5.5. The parts list is as follows.

R_1. 47-kilohm, 1-watt, 10%.
R_2. 2-kilohm potentiometer.

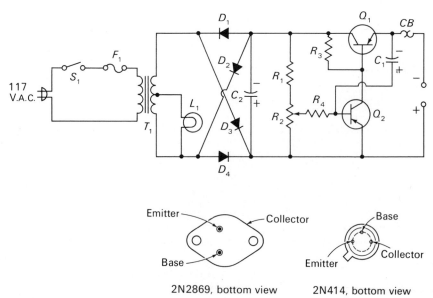

2N2869, bottom view   2N414, bottom view

Figure 5.5
Circuit diagram of variable low-voltage power supply.

R_3, R_4. 1.2-kilohm, 1-watt, 10%.

C_1. 1-microfarad, 25-volt, electrolytic.

C_2. 1,000-microfarad, 25-volt, electrolytic.

D_1, D_2, D_3, D_4. 1N1218.

Q_1. 2N2869.

Q_2. 2N414.

F_1. Fuse, 1-ampere.

CB. Circuit breaker, 1-ampere (Mallory type CB100, or equivalent).

T_1. 12.6-volt center-tap transformer (Stancor P-6134, or equivalent).

L_1. Pilot lamp, #47.

Miscellaneous: Vectorbord foundation, minibox, switch, red binding post, black binding post, fuse holder, push-in terminals, heat sink, hardware, wire.

Functional Description

A full-wave bridge rectifier (D_1, D_2, D_3, D_4) and filter capacitor (C_2) provide a nominal 20 volts to the control circuit (Q_2) and series-connected power transistor (Q_1). The control is achieved by varying the series resistance of Q_1. The resistance of Q_1 is varied by varying the base voltage of Q_2 through control R_2. Final filtering is achieved by applying the load ripple voltage to the base of Q_2 through coupling capacitor C_1. The ripple voltage is amplified by Q_2 and phase-inverted. It is then applied to the base of Q_1, where it is 180° out of phase with the original ripple voltage and cancels most of it out.

Construction Notes

Transistor Q_1 is a power transistor that must be mounted on a heat sink and electrically insulated from other components and the chassis. A suitable heat sink can be constructed from a piece of aluminum approximately $\frac{1}{4} \times 4 \times 3$ inches in size. Silicone compound and an insulating washer must be used between the transistor and the heat sink.

The value of resistor R_1 may require adjustment to provide the desired minimum output voltage from the supply. The minimum voltage should be approximately 0.2 volts, with R_2 turned down as far as possible.*

---

* For a detailed description of the above power-supply circuit, see *Electronics World.* 78, no. 4 (Oct. 1967), 79, "Variable Low-Voltage Power Supply," by Murray S. Rifkin.

VARIABLE LOW-VOLTAGE POWER SUPPLY (2-AMPERE)

Applications and Specifications

1. A transistor-regulated power supply that provides a continuous D.C. voltage from about 4 volts to 24 volts, and is suitable for operating digital programming circuits and relay circuits that need a well-regulated source of power.

2. Load regulation: Regulates up to 24 volts with a 2-ampere load, up to 6 volts with a 3-ampere load.

3. Ripple voltage: Negligible.

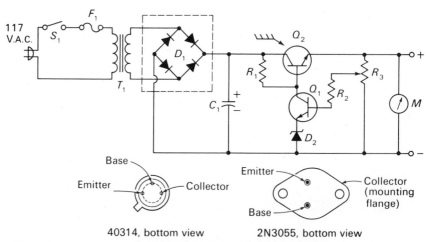

Figure 5.6
Circuit diagram of variable low-voltage power supply (2-ampere).

The circuit diagram is shown in Figure 5.6. The parts list is as follows.

R_1. 100-ohm, 10-watt, 10%.
R_2. 1-kilohm, ½-watt, 10%.
R_3. 1-kilohm, linear-taper potentiometer, 2-watt.
C_1. 25,000-microfarad, 50-volt, electrolytic.
Q_1. Type 40314.
Q_2. 2N3055.
D_1. Motorola type MDA962-3 rectifier bridge, or equivalent.
D_2. 1ZF3.9T10.

Heat sink. G.E. type for 1 TO-3 transistor.

Meter. Simpson model 1329: 4½-inch, 0- to 30-volt, D.C. volt-meter, or equivalent (optional).

T_1. Triad type F47V, or equivalent.

F_1. 2-ampere fuse.

S_1. SPST switch.

Miscellaneous: Binding posts, fuse holder, heat sink, teflon insulator, mounting hardware, chassis, wire.

Functional Description

Power transformer T_1, full-wave bridge rectifier, module D_1, and capacitor C_1 provide a nominal 24 volts to a series-regulated control circuit. Zener diode D_2 provides a constant nominal 3.9-volt potential to the emitter of transistor Q_1. When the supply output voltage tends to increase, so does the Q_1 base voltage, which lowers both the Q_1 collector current and the base current of Q_2. The resulting reduction in Q_2 collector current effectively counteracts the initial increase in the output voltage, providing instantaneous voltage regulation. The amount of Q_1 base voltage is controlled by potentiometer R_3, whose setting determines the value of the regulated output voltage. Resistor R_1 is selected to keep the zener diode D_2 in its saturated voltage region.

Construction Note

Transistor Q_2 must be mounted on a heat sink for adequate heat dissipation. The heat sink should be suitably insulated from the chassis as shown in Figure 3.14.

GENERAL-PURPOSE POWER SUPPLY (24-VOLT)

Applications and Specifications

1. A general-purpose power supply especially suited for relay, electro-mechanical counter, and recorder operation.

2. Maximum current rating: 2 amperes or 4 amperes (see Operation Note).

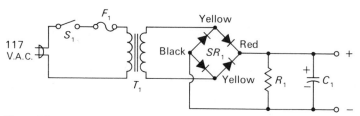

Figure 5.7
Circuit diagram of 24-volt, general-purpose power supply.

The circuit diagram is given in Figure 5.7. The parts list is as follows.

R_1. 250-ohm, 10-watt, 10%.

C_1. 50-microfarad, 50-volt, electrolytic.

SR_1. Selenium rectifier stack (International Rectifier type J29B5, or equivalent).

T_1. 24-volt filament transformer (Stancor P8357, or equivalent).

F_1. 2-ampere fuse.

Miscellaneous: SPST switch, fuse holder, binding posts, foundation, push-in terminals, hardware.

Functional Description

A 24-volt filament transformer T_1 provides current to bridge rectifier SR_1. The output of the rectifier is filtered through an RC circuit.

Construction Note

Layout of parts is not critical, but air should be able to circulate around the selenium rectifier stack.

Operation Note

This circuit can be made to provide up to 4 amperes by substituting a higher amperage 24-volt power transformer for the one specified in the parts list.

SIMPLE RELAY POWER SOURCE (28-VOLT)

Applications and Specifications

1. A 28-volt power source suitable for one or two 28-volt D.C. relays or solenoids commonly used in behavioral experimentation.

2. Power requirements: 117 volts A.C. at 150 milliamperes.

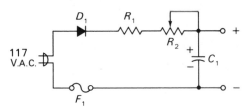

Figure 5.8
Simple 28-volt relay power supply.

The circuit diagram is given in Figure 5.8. The parts list is as follows.

R_1. 500-ohm, 10-watt, 10%.
R_2. 1,000-ohm, 10-watt rheostat (Ohmite type 0119, or equivalent).
C_1. 20-microfarad, 50-volt, electrolytic.
D_1. 150-milliampere, D.C., 25-volt selenium rectifier (IR type BIH, or equivalent).
F_1. ⅛-ampere fuse.
Miscellaneous: Circuit foundation, fuse holder, binding posts.

Functional Description

House power is half-wave rectified by D_1 and dropped in voltage through R_1 and R_2. It is filtered by C_1 and applied directly to the load.

Construction Notes

The selection of the correct resistance setting for R_2 depends on the resistance of the relay coil being used, which in general will be between 500 and 1,500 ohms. The correct setting of R_2 may be ascertained by placing a voltmeter across the relay coil and adjusting R_2 until the meter reads 28 volts.

If the relay chatters on application of power, the value of capacitor C_1 should be increased to provide greater filtering.

HIGH-VOLTAGE D.C. LABORATORY POWER SUPPLY

Applications and Specifications

1. A general-purpose power supply suited for shock circuits, lesion-maker circuits, and other uses where a well-filtered source of high-voltage D.C. is required.

2. Output: 800 volts D.C. at 175 milliamperes. The supply may be made to operate as a variable voltage source (see Operation Note).

3. Load regulation: 16%.

4. Ripple voltage: 1%.

The circuit diagram is given in Figure 5.9. The parts list is as follows.

C_1, C_2. 4-microfarad, 1,000-volt (Cornell-Dubilier 10040).

C_3, C_4. 20-microfarad, 600-volt, electrolytic.

C_5–C_{20}. .01-microfarad, 600-volt, ceramic.

CR_1. 16 G.E. type 1N1696 silicon rectifier diodes connected in groups of four.

CR_2. G.E. type 6RS20SP4B4 Thyrector diode (optional transient voltage suppressor).

F_1. 3AGC fuse, 3-ampere.

I_1. 120-volt, 6-watt pilot lamp.

L_1. $\frac{5}{25}$-henry choke, 175-milliampere (UTC S-30, or equivalent).

L_2. 10-henry choke, 175-milliampere (UTC S-29, or equivalent).

L_3. 30-henry choke, 25-milliampere (UTC S-25, or equivalent).

R_1. 50,000-ohm, 25-watt.

R_2. 110,000-ohm, 4-watt (two 220K, 2-watt resistors in parallel).

R_3–R_{18}. 250,000-ohm, $\frac{1}{2}$-watt.

S_1. DPST switch.

T_1. 200-milliampere transformer: primary, 120-volt A.C., 60 hps; secondary, 800-volt (Stancor PC-8412, or equivalent).

Miscellaneous: Chassis, binding posts, solder lugs, terminal strips, fuse holder, pilot lamp, lamp socket, wire.

Functional Description

A high voltage transformer (T_1) provides current to 16 silicon rectifiers, which form a bridge rectifier capable of handling the 800-volt transformer output. The output of the bridge rectifier is filtered by a choke input filter,

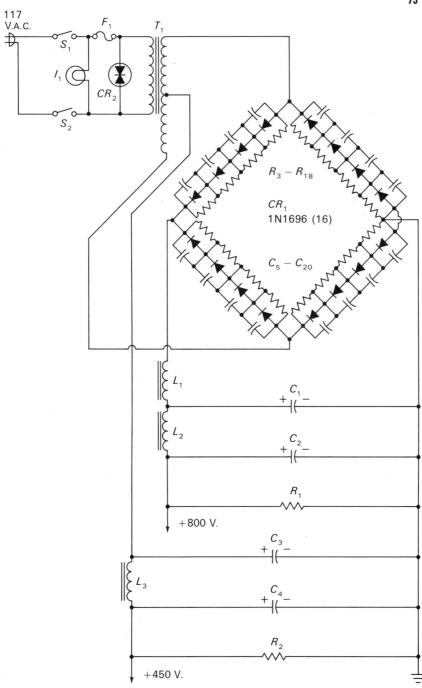

Figure 5.9
Circuit diagram of high-voltage D.C. laboratory supply.

L_1, and by the combination of C_1, L_2, and C_2. Resistor R_1 is a bleeder resistor that discharges the filter circuit when the circuit is deenergized, and also improves the load regulation. Capacitors C_5 through C_{20} and resistors R_3 through R_{18} protect the diodes from transient voltages and reverse resistance variations, respectively.

Construction Notes

The 16 silicon rectifier diodes, resistors, and capacitors may be mounted on four terminal strips with each terminal strip representing one leg of the bridge circuit.

A pilot light should be included in the circuit to indicate when the circuit is energized.

Operation Note

This supply may be made to operate as a variable voltage source by incorporating an adjustable transformer (Variac) in the line ahead of the step-up transformer.

Cautions

The 800-volt output of this supply could be dangerous or even lethal. Care should be exercised in the construction and use of the supply in order to avoid dangerous shocks.*

---

* For a detailed description of a similar power supply, see General Electric Company, *Silicon-Controlled Rectifier Hobby Manual* (Auburn, New York: General Electric Company Rectifier Components Department, 1963), pp. 23–25.

D.C.-TO-D.C. HIGH-VOLTAGE CONVERTER

Applications and Specifications

1. A converter that provides high-voltage D.C. from a low-voltage battery supply; suitable as a safe source of shock voltages.
2. Input: 3 volts D.C. (two 1½-volt D cells in series).
3. Output: Variable high voltage D.C., 1,000 volts at 25 microamperes maximum.

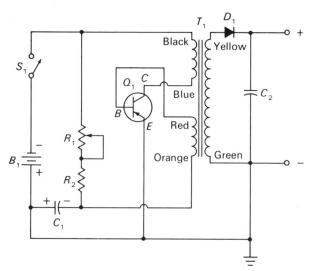

Figure 5.10
Circuit diagram of D.C.-to-D.C. high-voltage converter (courtesy of Triad Distributor Div.).

The circuit diagram is given in Figure 5.10. The parts list is as follows.

R_1. 10-kilohm, 2-watt, linear-taper potentiometer.
R_2. 1-kilohm, 1-watt, 10%.
C_1. 2.2-microfarad, 10-volt.
C_2. 0.02-microfarad, 1,600-volt, mica, ceramic, or oil-filled paper.
Q_1. 2N670.
D_1. High-voltage rectifier (International Rectifier T-35-HP).
T_1. Triad type TY-200X.

B_1. Two $1\frac{1}{2}$-volt D cells connected in series.

Miscellaneous: Switch, binding posts, circuit foundation, enclosure, hardware, wire.

Functional Description

Transistor Q_1 is employed in a blocking oscillator circuit. Feedback necessary to start and sustain oscillations is provided by transformer T_1. The output appears at the transformer secondary, where it is rectified and filtered by D_1 and C_2, respectively. The frequency of the oscillator is determined by the time constant of R_1, R_2, and C_1. Varying control R_1 effectively varies the output voltage.

Construction Note

Layout of parts is not critical, although all leads should be kept short and direct to reduce undesirable radiation.

Operation Notes

The output voltage is controlled by R_1, which should be adjusted after the circuit is connected to a load, and which should be set at maximum resistance before power is applied.

Transistor Q_1 and transformer T_1 may be damaged if any attempt is made to increase the output beyond 1,000 volts.

D.C.-TO-A.C. INVERTER (12 VOLTS D.C. TO 120 VOLTS A.C.)

Applications and Specifications

1. Converts 12-volt D.C. battery current (from an automobile storage battery) to 120-volt, 60 cycle A.C. (approximating a square wave).
2. Power output: 120 watts maximum.
3. Useful for operation of laboratory equipment in the field.
4. This type of inverter should not be used to power equipment that needs a precise sine-wave frequency.

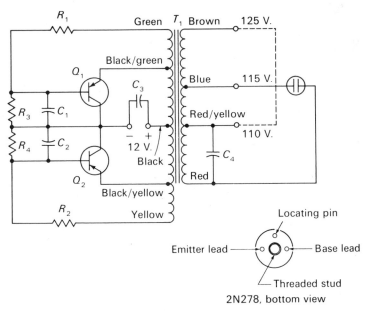

Figure 5.11
Circuit diagram of D.C.-to-A.C. inverter.

The circuit diagram is given in Figure 5.11. The parts list is as follows.

R_1, R_2. 2-ohm, 5-watt, wire-wound.
R_3, R_4. 200-ohm, 10-watt, wire-wound.
C_1, C_2. 2-microfarad, 25-volt, electrolytic.
C_3. 250-microfarad, 25-volt, electrolytic.
C_4. 1-microfarad, 200-volt.

Q_1, Q_2. 2N278.

Two Wakefield type NC421B natural convection coolers (heat sinks), or equivalent.

T_1. Triad type TY-75A.

Miscellaneous: 14-gauge wire, hook-up wire, case, outlet receptacle, battery clips, hardware.

Functional Description

This circuit is designed around transformer T_1, a specially wound inverter transformer. Transistors Q_1 and Q_2, together with their associated bias resistors and capacitors, form an oscillator circuit whose frequency is determined by the induced voltage in the transformer and by the input voltage. When the circuit is oscillating, the transistors act as switches, alternating in their conduction. The output voltage is induced in the T_1 secondary windings, which are tapped for 110 volts, 115 volts, or 125 volts. The capacitors in the circuit help filter the voltage spikes induced by the transformer.

Construction Notes

The transistors may be mounted directly to the heat sinks and the heat sinks mounted directly to the chassis (ground). For maximum heat dissipation, the heat sinks should be mounted on the outside surface of the inverter case.

Battery leads should be made from 14-gauge or larger wire, and should be kept as short as possible.

Operation Notes

Observe correct battery polarity.

To avoid transistor-damaging voltage transients, the load should be attached to the inverter with the load in its operating mode (i.e., switched on).

Of the three output voltages 115 volts is the most commonly used.

VOLTAGE DOUBLERS

Applications and Specifications

1. Voltage doublers approximately double r.m.s. supply or transformer-coupled A.C. voltages while providing filtered direct current.

2. Two circuits are presented: a conventional half-wave doubler, and a voltage quadrupler made from two doubler circuits.

3. Regulating characteristics: of doubler circuit; approximately 300 volts at 0.1 ampere and 235 volts at 0.5 ampere; of quadrupler circuit, approximately 600 volts at 25 milliamperes and 500 volts at 0.2 ampere.

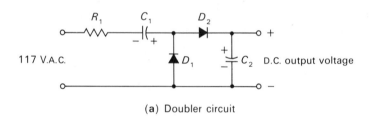

(a) Doubler circuit

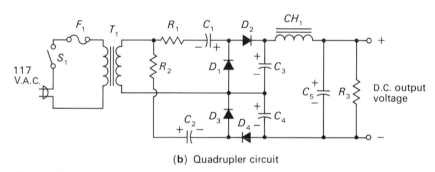

(b) Quadrupler circuit

Figure 5.12
Circuit diagrams of: (a) voltage doubler; and (b) voltage quadrupler.

The circuit diagrams are given in Figure 5.12. The parts list for the voltage doubler is as follows.

R_1. 5.6-ohm, 10-watt.
C_1, C_2. 100-microfarad, 450-volt, electrolytic.
D_1, D_2. 1N1763.

The parts needed for the voltage quadrupler are as follows.

C_1, C_2. 100-microfarad, 150-volt, electrolytic.
C_3, C_4. 80-microfarad, 450-volt, electrolytic.
C_5. 20-microfarad, 600-volt, electrolytic.
CH_1. 8.5-henry, 200-milliampere choke.
R_1, R_2. 10-ohm, 10-watt, wire-wound.
R_3. 50,000-ohm, 10-watt.
D_1, D_2, D_3, D_4. 1N1763.
T_1. Power transformer, 117-volt secondary at 0.3 ampere (Triad type N-51X, or equivalent).
F_1. 2-ampere fuse.
S_1. SPST switch.
Miscellaneous (for both circuits): Terminal strips, chassis, pilot light, hardware, wire.

Functional Description

The voltage-doubler circuit is known as a half-wave type. Capacitor C_1 is charged on one half of the input cycle; capacitor C_2 is charged on the other half of the input cycle. Whereas C_1 is charged directly by current-limiting resistor R_1 and the line voltage, C_2 is charged by both the line voltage and the charge on C_1. The diodes direct the current to the appropriate capacitor, as well as rectify the alternating current. During the negative half-cycle, D_1 permits C_1 to be charged, while D_2 blocks electricity from C_2. During the positive half-cycle, D_2 permits the charge from C_1 to charge C_2, while D_1 blocks electricity from charging C_1. The voltage appearing across capacitor C_2 is approximately twice that of the input voltage.

The voltage quadrupler is essentially two voltage-doubler circuits that alternately provide their individual doubled outputs to a common circuit. In addition to the two doublers, the quadrupler circuit includes an isolation transformer (T_1) that protects against shock, and a choke (CH_1) and capacitor (C_5) that reduces the ripple content to a low value.

Construction Note

Observe correct polarity of diodes and electrolytic capacitors.

Cautions

The doubler circuit diagram is shown with its input directly across the line. Although the circuit will function as shown, it is extremely dangerous to operate it in this mode, since it is possible for the high-voltage output to be 117 volts A.C. above the ground potential, presenting a serious shock hazard. An isolation transformer, such as the one shown for the quadrupler circuit, should be used.

A pilot light should be used to indicate when power is provided to the circuits. A neon-type indicator containing a built-in current-limiting resistor, such as the Tineon type 36N-2311, placed across the transformer primary windings, will serve as such a pilot lamp.

REGULATED STORAGE-BATTERY CHARGER

Applications and Specifications

1. Provides automatic, metered charging for automobile storage batteries. The charger automatically turns itself off when the battery is charged to a preset level.

2. Power requirements: 110 volts A.C., 2 amperes maximum.

3. Output charging voltage: 12 volts D.C.

4. Output charging current: 0-10 amperes.

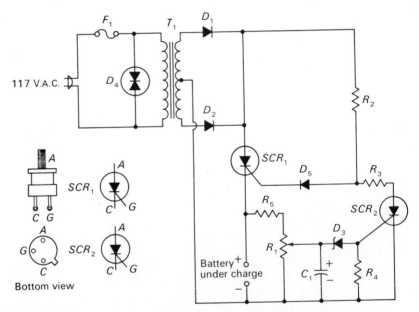

Figure 5.13
Circuit diagram of regulated storage-battery charger.

The circuit diagram is given in Figure 5.13. The parts list is as follows.

R_1. 500-ohm, 2-watt, linear potentiometer.
R_2, R_3. 27-ohm, 3-watt.
R_4. 1,000-ohm, ½-watt.
R_5. 47-ohm, 1-watt.

C_1. 100-microfarad, 25-volt, electrolytic.

D_1, D_2. 15-ampere, 50-volt silicon rectifier (GE-X4).

D_3. 8.2-volt zener diode, 1-watt (GE-X11).

D_4. GE thyrector type 6R20SP4B4.

D_5. 1N1692.

SCR_1. Silicon-controlled rectifier (GE-X3).

SCR_2. Silicon-controlled rectifier (GE-X5).

T_1. Power transformer: primary, 117 volts A.C.; secondary, 24 volts, CT (Triad F41X, UTC-FT10, or equivalent).

F_1. 2-ampere fuse.

Miscellaneous: Heat sinks (Thermalloy type 6123B, or equivalent), stand-off insulators, binding posts, fuse holder, line cord, wire, hardware.

Functional Description

The charger circuit consists of: a center-tapped transformer, T_1; full-wave rectifier circuit, D_1 and D_2; a switching element, SCR_1; and a regulating circuit, SCR_2, R_1, C_1, and D_3. The regulating circuit provides the gating signal for SCR_1. The battery under charge provides a voltage signal to the gate of SCR_2, according to the setting of R_1 and the charge of C_1. As the battery voltage rises to its fully charged state and the charge on C_1 increases, the zener diode (D_3) conducts, turning on SCR_2, which being in series with R_2 and R_3, forms a voltage divider and prevents a positive signal from turning on SCR_1. With SCR_1 turned off, the battery is effectively separated from the charging source.

Construction Notes

Two heat sinks are used to mount the diodes, D_1 and D_2, and SCR_1. If the diodes are bolted directly to the heat sinks, the heat sinks should be electrically isolated from the chassis by stand-off insulators; SCR_1 may be insulated from the heat sink by insulated washers and silicon grease. After all components are mounted on the chassis, a multimeter should be used to check for short circuits between the diodes and chassis.

Thyrector D_4 is a special semiconductor that suppresses transient voltages across the power line and keeps them from damaging the other semiconductor circuit elements. Although the circuit will function adequately without D_4, the latter is recommended if protection against power-line surge voltages is necessary.

Charger output terminals should be adequately labeled to prevent attaching of the battery to the wrong charger terminals.

300148

Operation Notes

The charger is set up initially on a freshly charged battery (as checked with a hydrometer). With the charger attached to the battery, control R_1 is rotated until the meter reads zero.

A large load (3 ohms, 50 watts) attached to the battery will cause the charger to operate (indicating approximately 4 amperes), and if the load is removed, the charging rate should taper off and return the meter to zero, indicating a fully charged battery.

Once set up, the charger will automatically charge a discharged battery to the predetermined level and then maintain that level.*

---

* For further information about this circuit, see General Electric Company, *Silicon-Controlled Rectifier Hobby Manual* (Auburn, New York: General Electric Company Rectifier Components Department, 1963), pp. 27–28. For detailed design information about regulated battery chargers, see "Regulated Battery Chargers Using Silicon Controlled Rectifier," by D. R. Grotham, General Electric Company Application Note 200.33.

6

Sensing Circuits

Sensing circuits are commonly used in behavioral experimentation to indicate responses that one wishes to control or record. The term sensing circuitry, as used in this chapter, refers to electrical circuits composed of some form of activity sensor and of an initial control stage. A sensor is any device that detects activity in one system and delivers information about this activity to another system. The initial control stage provides a response indication, or signal, which can be used later in a final control stage for further processing or can be recorded.

Experimental designs may be conveniently divided into stages or circuits. Figure 6.1 is a block diagram of a typical experimental system. The control circuitry operates with the information from the sensing circuitry, and in behavioral experimentation typically controls the stimulus and reinforcement contingencies. All the stages shown in Figure 6.1 are controlled by the programming circuitry. The programming circuitry and the control circuitry often function together as a single circuit. This chapter will concern itself exclusively with sensing circuitry; succeeding chapters will discuss control, programming, and recording circuits.

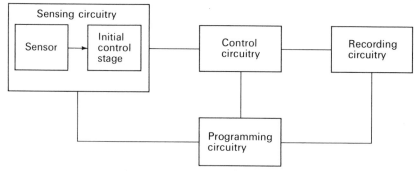

Figure 6.1
Block diagram of an experimental system.

SENSORS

In behavioral experimentation the subject often acts directly as a sensing element in a circuit by responding to stimuli. Perhaps the simplest example of the subject functioning as a sensor is in its activation of switches that control electrical circuits, which may be used to energize a control or recording circuit.

A more complex example is the activation of a control relay by the subject's proximity to another sensing element, as happens with a body-capacitor electronic relay. In this example the body capacitance of the subject detunes an oscillator circuit that is usually stable and maintains a control relay in an inoperative state. When the circuit is detuned, the relay is activated, responding to the proximity of a subject to the oscillator circuit. When the subject leaves the vicinity of the oscillator circuit, the tuning again stabilizes and the relay deenergizes.

In addition to using the subject himself, the experimenter may use other devices as sensing elements in an electrical circuit. These devices, known as transducers, use photic, chemical, thermal, piezoelectric, or mechanical energy to affect current flow in an electrical circuit. The most commonly used transducers either generate voltage or change resistance in a circuit when excited by enough energy. Transducers may also control circuits directly; an example is the photo transistor, a PNPN switch triggered by light energy. Table 6.1 lists several transducers commonly used in electronic sensing circuits.

The circuits described in this chapter may be used in experimental set-ups that need to collect precise data from low-level transducer signals. These signals are often of such small magnitude (microamperes) that they must be amplified before they can be used to provide useful information or to control other experimental parameters. The transducer is often integrated into an initial control stage, which may consist of a sensitive

Table 6.1
Transducers Commonly Used in Electronic Sensing Circuits.

Type of energy	Example	Voltage generator (V) Resistive element (R)	Material
Photic	Photovoltaic Cell	V	Selenium, Silicon Cadmium Sulfide
	Photoconductive Cell	R	
Thermal	Thermistor	R	
	Dissimilar Metals	V	Copper, Constantian
Chemical	Storage Battery	V	Nickle-cadmium, or lead and sulphuric acid
Mechanical	Electric generator	V	Magnetic field, moving coil
Piezoelectric	Crystal Microphone	V	Quartz crystal

relay, an electronic amplifier, or some other control element that effectively amplifies the low-level transducer signal.

Other circuits in the initial control stage are needed to shape the sensor signal. A pulse former is an example of a shaping circuit. It is used to provide signals of some standard duration from sensor signals that may be of varying duration. Thus signals from the depression of a lever or the pecking of a key, which might vary greatly in duration, can be standardized with a pulse-former circuit into uniform pulses of fixed duration. The standardized pulses may then provide uniform signals to a control or recording stage that might function erratically given nonuniform input signals.

INITIAL CONTROL CIRCUITRY

The initial control circuitry is connected directly to the sensor, and together they form the complete sensing circuit. The most common initial control stage used in behavioral experimentation is the electro-mechanical relay. A relay may be considered an electrically operated switch. Basically it consists of an energizing coil through which current flows to produce a magnetic field that attracts a spring-loaded armature. Contact points on the armature and on the coil foundation enable switch-type contacts to be made that are electrically isolated from the coil. Relays are available in a variety of contact configurations and coil types. Common contact configurations include SPDT (shown schematically in Figure 6.2), DPDT, and 4PDT. Coil types may be obtained for both A.C. and D.C. service, with voltage ratings from 6 to 400 volts. Appendix E is a listing of useful relay circuits with brief circuit descriptions. In addition to the

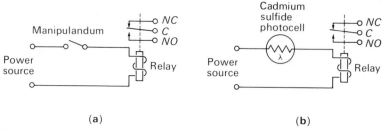

(a)   (b)

Figure 6.2
Initial control circuits using relays: **(a)** mechanical switch; **(b)** resistive sensors.

common relay configurations, a large variety of specially constructed relays are available. These relays perform different types of switching functions, including stepping, sequencing, and latching.

As is shown in Figure 6.2, it is sometimes possible to attach a relay directly to a sensor in a simple electric circuit. The switch shown in Figure 6.2, a, may be any type of switching device that can deliver operating power to the relay. Figure 6.2, b, shows a resistive sensor used as a switching element to control the operation of a relay. The cadmium sulfide photocell responds electrically to changes in the light falling on its surface. In darkness, the resistance, as measured across the cell, may be as high as several megohms. When light strikes the cell, the resistance may drop to as low as several hundred ohms, depending on the type of cell and on how much light reaches the cell. When the cell resistance is high, the relay will not receive enough current to close its contacts. With the cell resistance lowered, enough current will flow to activate the relay. Figure 6.3 illustrates how such a photocell sensor might be used to detect the motion of a subject that interrupts the photobeam. The lamp and the lens assembly direct a narrow beam of light onto the photocell. As long as the cell is illuminated, the control relay remains energized. When a subject blocks the light, the relay deenergizes, its contacts close, and the control circuitry is activated. The potentiometer controls sensitivity by limiting the current through the photocell and relay. The sensitivity of this circuit is poor. Typically, 50 to 100 foot-candles of light must shine on the photocell to activate the relay. For proper operation, the distance between the photocell and the light source must be small, and the environment surrounding the cell must be kept dark. To improve the sensitivity of this circuit, a transistor amplifier may be used to amplify relatively small changes in the resistance of the photocell into enough power to operate a control relay. Circuits using transistor amplifiers to amplify the output of photocells and other resistive transducers are presented in this chapter. The sensitivity of a photocell circuit may be improved by a transistor amplifier to where the relay can be operated by less than one foot-candle of light.

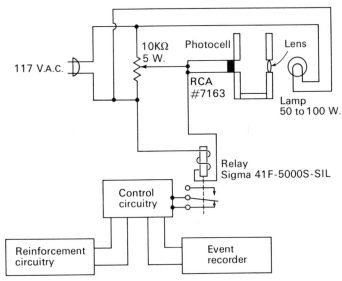

Figure 6.3
Block diagram of simple photoelectric relay and control circuit.

In the circuits described above, a resistive sensor is used to activate a relay in an initial control stage. Circuits operating from voltage sensors may be similarly constructed, using the current generated by the sensor either to control a sensitive relay directly or to provide a bias voltage for operating a transistor amplifier.

The silicon-controlled rectifier (SCR) is a sensitive electronic switch suitable for use with sensors and mechanical switches. The SCR is a solid-state device that, when pulsed through its gate lead, functions like a conventional rectifier. Once the device is conducting, the gate lead no longer has control over current flow. To switch the SCR to a conducting state, a relatively low-level voltage, positive with respect to the SCR cathode, is applied to the gate. A momentary trigger pulse of a few microseconds' duration is enough to initiate conduction. Conduction through the SCR continues until its anode voltage drops to zero, at which time the SCR is again controlled by the gate. Either voltage sensors or switches may be used to provide trigger signals. Representative switching circuits using SCRs are included in this chapter.

ELECTRIC THERMISTOR THERMOMETERS

Applications and Specifications

1. Two temperature-measuring circuits are described, each using a thermistor as the sensing element.

2. Both circuits will enable temperature measurements to be made between approximately -18 and $50°C$.

3. Accuracy of temperature measurements: circuit 1, $\pm15\%$ (approximate); circuit 2, $\pm2\%$ (approximate). The accuracy of both circuits depends on the accuracy of initial calibration, the temperature range to be measured, and thermistor tolerances.

4. Power requirements: circuit 1, two 1.4-volt mercury batteries with no adjustment for battery deterioration; circuit 2, three 1.4-volt mercury batteries with compensation adjustment for battery deterioration.

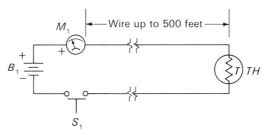

Figure 6.4
Circuit diagram of electric
thermistor thermometer (1).

The circuit diagram for circuit 1 is given in Figure 6.4. The parts list for circuit 1 is as follows.

> M_1. 100-microammeter (Simpson #1227, or equivalent).
> TH. Thermistor, Fenwal type EMC4.
> S_1. SPST, push-button.
> B_1. 1.4-volt mercury cells (two).

The circuit diagram for circuit 2 is given in Figure 6.5. The parts list is as follows.

> R_1. 1-kilohm, $\frac{1}{2}$-watt, 10%.
> R_2. 500-ohm, 2-watt, wire-wound control.

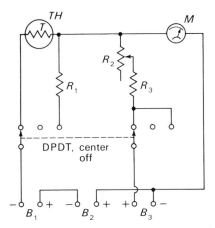

Figure 6.5
Circuit diagram of electric
thermistor thermometer (2).

R_3. 600-ohm, ½-watt, 10%.
TH. YSI thermistor, Type 44003.
M. 200-microampere full-scale meter.
B_1, B_2, B_3. 1.4-volt mercury penlight batteries.
S_1. DPDT, center-off.
Miscellaneous (for both circuits): Battery holders, wire leads for
thermistors, chassis, case, wire, hardware.

Functional Description of Circuit 1

This simple circuit is useful for monitoring temperatures between −18
to 50°C. It can be calibrated against a standard thermometer and will
read temperature changes to within 15 or 20 per cent accuracy, depending
on the meter movement and scale size. As with all thermistor measuring
devices, it will respond within milliseconds to changes in the temperature
of a bathing fluid, but will take about 30 seconds to respond to changes in
air temperature. The circuit contains two long-life mercury cells, which
will give reliable, accurate readings for up to six months when used
continuously.

The circuit consists of a thermistor wired in series with a sensitive
microammeter. When the push-button switch is pressed, current flows
through the thermistor and is read directly by the meter. At room temperature the total current is about 20 microamperes. This current will tend to
heat the thermistor with a power in the microwatt range, and will result in
a temperature rise of only .05°F. above room temperature, which can be
ignored when the meter is calibrated against a standard thermometer.

Functional Description of Circuit 2

This circuit can precisely measure temperatures in a variety of ranges. With careful initial calibration, accuracies within 2 per cent are possible over a range of 30°F. As shown in the circuit diagram, a thermistor, two batteries (B_1 and B_2), and a meter are connected in series. The meter can also have a reverse current introduced to it through an adjustable voltage control (R_2), a fixed resistor (R_3), and a third source of current (B_3). A DPDT switch with a center-off position is used to switch a fixed resistor (R_1) into the circuit for calibration purposes. In the calibration position, the thermistor is removed from the circuit and fixed resistor R_1 is used to simulate a reference temperature; with the value of R_1 at 1,000 ohms, the simulated temperature is 25°C., and control R_2 is then used to adjust the meter to read 25°C. This is the only adjustment necessary, and when the switch is placed in the read position, the meter reads directly the temperature sensed by the thermistor.

Construction Notes

The thermistors used in both circuits may be located up to 500 feet from the meter, using two-conductor cable.

Both circuits may be calibrated against precise reference temperatures. Both the thermistor probe and the reference thermometer should be immersed in an oil or water bath and the temperature raised until the thermometer reads within the operating range of the thermistor. Temperature settings may then be marked directly on the face of the meter or a chart of meter readings versus temperature may be constructed.

An alternate method of calibration, which involves comparison of resistance and temperature readings as specified by the thermistor manufacturer, is particularly adaptable to circuit 2, because the thermistor is manufactured to be within 1 per cent of its specifications. Using this method, precision resistors are substituted for the thermistor, and meter readings are taken. The temperature corresponding to a particular resistance value is found on the resistance-temperature curve that accompanies the thermistor. The meter reading may then be listed as corresponding to a particular temperature value, or the meter scale may be carefully removed and a new scale (reading directly in degrees) may be substituted.

A calibration mark should be made on the meter face of circuit 2, as follows. With switch S_1 in the "calibrate" position, rotate control R_2 until the meter needle falls close to the low end of the scale (corresponding to approximately 40 microamperes). Place a reference calibration mark on the meter face at this point. This mark allows for periodic meter calibration. The calibration control R_2 permits the meter to be referred to the calibration mark despite changes in battery voltage due to aging. It greatly extends the useful operating life and accuracy of the circuit.

The operating range of circuit 2 is approximately between 20 and 33°C. This range may be changed by substituting other values for resistor R_1 and control R_2. If, for example, it is desired to include measurements down to 4°C., then resistor R_1 should be replaced with one having the resistance shown on the thermistor calibration card for a reading at or near 4°C. The value of the zero reference control R_2 should be suitably changed to enable the meter to be accurately zeroed. In addition, if it is desired to extend the entire operating range of the circuit, a 1-microampere meter may be employed, or shunts may be used in parallel with the specified meter.*

---

* A slightly modified version of the thermometer in circuit 2 was originally described in Tony Karp, "Electronic Darkroom Thermometer," *Radio-Electronics*, 34 (Oct. 1963), 60, © Gernsback Publications, Inc., 1963. A thermistor manual describing thermistor circuits and their operation is available from Fenwal Electronics, 63 Fountain Street, Framingham, Mass.

SENSITIVE LIGHT METER

Applications and Specifications

1. Provides accurate measurements of light intensity from surroundings approaching darkness to normal daylight. Accuracy depends on accuracy of the standard light meter used to calibrate the meter.

2. Measures light levels in three ranges, from approximately 0.02 foot-candles to 20 foot-candles.

3. Provides a narrow "field of vision" for accurately measuring variations of light intensity over small areas.

4. Built-in battery condition indicator.

5. Power requirements: 1.5-volt penlight battery.

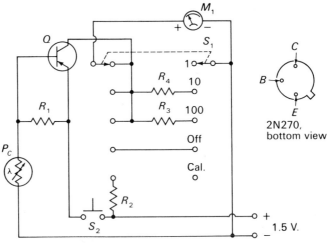

Figure 6.6
Circuit diagram of sensitive light meter.

The circuit diagram is given in Figure 6.6. The parts list is as follows.

M. 1-milliampere meter, 100 ohms resistance (Lafayette 99 R 5052).

Q. 2N270 (RCA).

PC. Cadmium-sulfide photocell (Lafayette 99 R 6321).

R_1. 15,000-ohm, $\frac{1}{2}$-watt, 10%.

R_2. 1,800-ohm, $\frac{1}{2}$-watt, 10%.

R_3. 1-ohm, 5%, wire-wound (IRC type BWH, or equivalent).

R_4. 10-ohm, 5%, wire-wound (IRC type BWH, or equivalent).

S_1. Subminiature rotary switch, 2 poles, 5 positions (Lafayette 99 R 6164, or equivalent).

S_2. Miniature momentary push-button, SPST, normally open (Lafayette 99 R 6218, or equivalent).

B. 1.5-volt carbon-zinc or mercury battery, penlight size.

Miscellaneous: Battery holder, case, tube for photocell tunnel, knob, wire, hardware.

Functional Description

A resistor (R_1) and cadmium-sulfide cell (PC) form a voltage divider, which regulates the base bias for a transistor (Q). In total darkness, the resistance of PC is very high (in the megohm range); in direct sunlight its resistance drops to several thousand ohms. When the resistance of PC is high, the current through the collector-to-emitter circuit of Q is small. When the resistance of PC is low, the current through the collector-to-emitter circuit of Q is large. The collector-to-emitter current flows through a meter that reads it directly; the reading is proportional to the intensity of the light shining on PC. Two shunts can be switched into the meter circuit. These shunts are used as range switches that effectively compensate for increased light levels shining on PC. A five-position rotary switch (S_1) is used to shunt the meter; remove the shunts when measuring dim light; check battery condition by introducing a fixed reference resistor (R_2) directly across the meter; disconnect the meter from the battery in the off position; and short circuit the meter movement to reduce the possibility of meter damage. A push-button switch (S_2) is in series with the emitter of Q and the meter. This switch prevents possible damage to the meter movement when S_1 is on the wrong range position.

Construction Notes

The photocell should be housed in a tunnel constructed from plastic or metal tubing and measuring $\frac{1}{2}$ inch in inside diameter and $\frac{1}{2}$ to $\frac{3}{4}$ inch in length. Paint the inside and outside of the tunnel flat black, and glue the photocell to the inside of the tunnel with epoxy glue, which can be used to fasten the entire assembly to the case as shown in Figure 6.7. Make sure that stray light cannot reach the photocell; seal the rear of the photocell against any light that may enter through holes in the meter case.

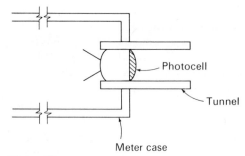

Figure 6.7
Mounting the photocell in a tunnel.

Operation Notes

When the range switch is in the calibration position, the meter will indicate battery condition. If the meter reads below 0.7 milliamperes, the battery should be replaced. In normal operation the battery should last about six months before replacement is necessary.

A calibration box may be constructed by painting a suitable container flat black and placing the constructed meter and a standard reference meter in the box next to each other. Suspend a 25-watt lamp above the box and connect it to a variable voltage source, such as a variable transformer (Variac), as shown in Figure 6.8. Aim both meters at the center of the bulb's brightest spot, and compare the readings of the meters at various bulb intensities (set by the variable transformer). Since most standard

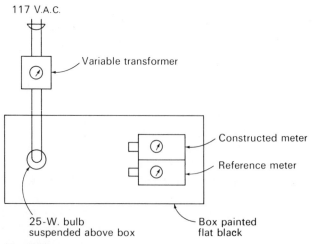

Figure 6.8
Calibration procedure.

light meters are calibrated in foot-candles, it is possible to construct a graph of foot-candles (as measured by the constructed meter). It should only be necessary to calibrate the constructed meter using the 100× range, since the other two ranges are multiples of it.

For photographic use, correct exposures may be calculated from the formula $FC = f^2/TS$, where FC is foot-candles, T is the exposure time in seconds, S is the ASA speed rating of the film, and f is the lens aperture (lens stop). Although this formula is only a rough approximation of the actual exposure calculation, it is adequate for most commercially available film.

In actual operation, the meter is aimed at the object to be measured, the range switch is set to the appropriate scale, and the push-button is depressed. The meter reading in milliamperes is then compared with the graph of foot-candles versus milliamperes. This graph may be glued to the rear of the meter for convenient reference.

Cautions

Set the range switch to the off position when the meter is not being used, since the meter movement is damped in this position.

If the meter needle reads too high when the push-button switch is depressed, release the button immediately and reset the range switch to the correct range.*

---

* For further information about this circuit, see Leon A. Wortman, "Build This Ultrasensitive Light Meter," *Radio-Electronics*, 37 (Feb. 1966), 50–52, © Gernsback Publications, Inc., 1966.

SIMPLE PHOTOELECTRIC RELAY

Applications and Specifications

1. A simple photoelectric relay control that responds to changes in light intensity reaching the photocell.
2. Sensitivity: responds to over-all changes in illumination from dark (night) to light (daylight). The sensitivity can be substantially improved by focusing a light source through a lens directed at the photocell; the relay can then function as a motion detector, responding when the beam of light is broken.
3. Operating power: 117 volts A.C.
4. Relay contact rating: 1 ampere.

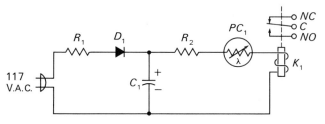

Figure 6.9
Circuit diagram of simple photoelectric relay.

The circuit diagram is given in Figure 6.9. The parts list is as follows.

R_1. 15-ohm, ½-watt.
R_2. 10,000-ohm, 2-watt.
C_1. 4-microfarad, 250-volt, electrolytic.
D_1. 130-microampere, 400-PIV rectifier (RCA 40265, or equivalent).
PC_1. Sigma 5HC2 photocell, 0.6-watt, 300-volt (or equivalent).
K_1. 24-volt D.C. relay, 1-ampere contacts.
Miscellaneous: Line plug, output terminals, case, wire, hardware.

Functional Description

The circuit consists of a D.C. power supply made from current limiter R_1, half-wave rectifier D_1, and filter capacitor C_1. The cadmium-sulfide photocell PC_1 is placed in series with control relay K_1 and current limiter R_2. When light falls on the photocell, its resistance drops and the relay is activated. In darkness, the cell resistance rises and the relay is deenergized.

Construction Notes

Since this is a line-operated circuit, take care to prevent short circuits. The case should not be grounded to the circuit. Observe correct polarity of diode and electrolytic capacitor.

Operation Note

Resistor R_2 affects the sensitivity of the circuit and its value may be changed slightly for the particular application.

SENSITIVE PHOTOELECTRIC RELAY

Applications and Specifications

1. Sensing of motion past photocell by interruption of a light beam directed at the photocell.

2. Automatic control of lighting; automatically turns lights on when no light is present.

3. Sensitivity using beam from a flashlight or similar light source: normal background lighting, 30 feet; total darkness background, 60 feet.

4. Operating power: 6-volt battery or D.C. power supply.

5. Minimum relay-coil operating current: 4 milliamperes.

6. Relay coil contact rating: 2 amperes at 115 volts A.C.

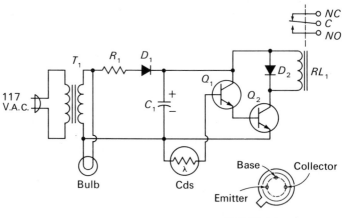

Figure 6.10
Circuit diagram of sensitive photoelectric relay.

The circuit diagram is given in Figure 6.10. The parts list is as follows.

C_1, C_2. 500-microfarad, 15-volt, electrolytic.

D_1, D_2. 1N2069.

Q_1, Q_2. 2N1302.

CdS. Cadmium-sulfide photocell (Allied Radio type 8E931, or equivalent).

RL_1. 1,000-ohm relay coil (Sigma 4F-1000 S-SIL, or equivalent).

Bulb. GE #44; 6.3-volt, 0.25-ampere (or equivalent).

T_1. 6.3-volt, A.C. filament transformer (Stancor P-6465, or equivalent).

Miscellaneous: Case, circuit foundation, bulb socket, lenses, lens holders, transistor sockets, wire, hardware.

Functional Description

The change in resistance of photoresistor CdS (in response to the light falling on the cell) causes the base voltage of transistor Q_1 to rise and enables the transistor to conduct. This conduction in turn causes transistor Q_2 to amplify the signal and actuate a sensitive relay. Diode D_2 protects the transistors from damaging transient voltages.

A simple 6-volt power supply (T_1, D_1, C_1, and R_1) provides operating power for the circuit. The transformer secondary also provides power for the pilot lamp that is the light source for CdS.

Construction Notes

For reliable operation, the photoelectric cell should be shielded from surrounding illumination by cementing it into a plastic or metal tube painted flat black. This tube can then be aimed at the light source that is to be interrupted. Additional sensitivity may be obtained by using a lens to concentrate the light from the light source onto the photocell.

BODY-CAPACITOR PROXIMITY RELAY

Applications and Specifications

1. Senses the movement of subjects past a detector plate, causing a relay to be activated when the subject is close to the detector.

2. Control relay and detector plate may be located remotely from the body-capacitor circuit.

3. Sensitivity: Performs reliably up to four inches from the detector plate for small subjects (laboratory rats) and up to twelve inches from the detector plate for large subjects (humans).

4. Power requirements: 6- or 12-volt D.C. battery or power supply. Current: 5 milliamperes with control relay open; about 15 milliamperes with the relay activated.

The circuit diagram is given in Figure 6.11. The parts list is as follows.

R_1. 470,000-ohm, ½-watt.
R_2. 10,000-ohm, ½-watt.
R_3. 680-ohm, ½-watt.
R_4. 22,000-ohm, ½-watt.
R_5. 330,000-ohm, ½-watt.
R_6. 2,000-ohm, ½-watt.
R_7. 2,000-ohm potentiometer.
R_8. 47-ohm, ½-watt.
C_1. 50-p.f., mica.
C_2. 3-to-30-p.f. trimmer.
C_3, C_4, C_5. 0.01-microfarad, paper.
C_6. 0.05-microfarad, paper.
Q_1, Q_2, Q_3. 2N1371.
Q_4, Q_5. 2N3708.
D_1, D_2. 1N34A crystal diode.
L_1. Ferrite core coil (J. W. Miller 4506, or equivalent).
L_2. R.f. choke, 2.5-millihenry.
K. Relay, 1,000-ohm D.C. coil (Sigma 4F-1000/S-SIL, or equivalent).
S_1. SPST, toggle.
S_2. DPDT, toggle.
Miscellaneous: Shielded cable (RG59), aluminum or copper sensing plate, battery, circuit foundation, terminals, case, hardware.

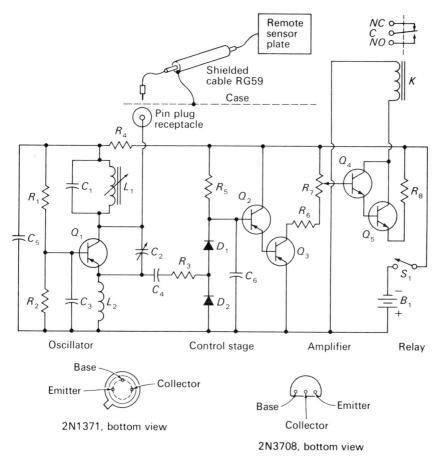

Figure 6.11
Circuit diagram of proximity detector.

Functional Description

The circuit is known as a loaded oscillator, because a change in the external dielectric of capacitance, such as that caused by the proximity of an animal, loads an oscillator circuit of which the external capacitance is a part. This loading effectively stops it from oscillating, and relay K operates, closing its NO contact, which may operate a recorder or other equipment. When the animal leaves the vicinity of the sensor, oscillations resume, and the relay contacts open. The circuit consists of an otherwise stable radio-frequency oscillator (approximately 4 megahertz), a control stage, an amplifier, and relay K. The control stage consists of a voltage doubler rectifier (D_1, D_2, C_4, and C_6). The output of the voltage doubler is applied to the base of transistor Q_2. With the oscillator stage operating, the output of the voltage doubler will cause Q_2 to be cut off. When oscillations cease (in the presence of a subject), Q_2 is biased "on" causing Q_3 to conduct. A second amplifier stage (Q_4, Q_5) provides optional added sensitivity and enables the sensitive control relay to operate. Base bias control R_7 functions as a sensitivity control, effectively regulating the conduction of Q_4.

Construction Notes

Location of components is not critical, except that L_1 and L_2 should be kept fairly far apart (at least 3 inches) to keep them from interacting.

The sensing plate is made of 1/16-inch aluminum and may be between 3 and 24 inches in diameter. The sensing cable is attached to the plate by a machine screw, solder lug, lock washer, and nut. It may vary in length, depending upon the body capacitance of the subject being sensed.

The control relay may be located up to 100 feet from the control circuit.

Operation Notes

The body capacitor is adjusted by setting sensitivity control R_7 for minimum resistance and then rotating capacitor C_2 clockwise using a nonmetallic screwdriver. At a certain point, relay K should close. Capacitor C_2 is then rotated counterclockwise until relay K just opens. If relay K does not open, control R_7 should be rotated until the relay opens. There is a certain amount of interaction between R_7 and C_2, but once C_2 is set, sensitivity can be controlled by R_7 exclusively. This completes the adjustments, and the relay should now energize when a subject approaches the sensing plate. These adjustments should be made only after the capacitor plate is established in its sensing location.

The body-capacitor relay will operate in moist soil if the sensing plate is encapsulated in plastic. It will not operate in wet soil, in liquids, or in close proximity to large metal objects.

The circuit may be modified to operate at higher voltages (18 to 24 volts D.C.) by eliminating R_6, R_7, Q_4, and Q_5, and by connecting a 5,000-ohm relay coil directly to points A and B (see circuit diagram). The sensitivity of the circuit is somewhat reduced when these modifications are made, although the reliability is the same.*

---

* This circuit was modified from one described in P. J. Shields, "Proximity Detectors and Metal Locators," *Howard W. Sams Photofact Publication*, PDS-1 (1965). The body-capacitor is also described in Mitchell Zucker and Walter E. Howard, "A Transistorized Body-Capacitance Relay for Ecobehavioral Studies," *Animal Behavior*, 16 (1968), 65–66.

ULTRASONIC PROXIMITY DETECTOR

Applications

1. Detection of moving objects near a 25-kilohertz note source. This circuit will sense the presence of laboratory animals 6 inches in length or larger.

2. Detector receiver can be located up to 50 feet from the transmitter and provides a narrow sensing beam for monitoring location or motion.

3. Detector is unaffected by associated stimuli, such as light or shock, if the stimuli are introduced through independent circuitry.

The circuit diagram is given in Figure 6.12. The parts list is as follows.

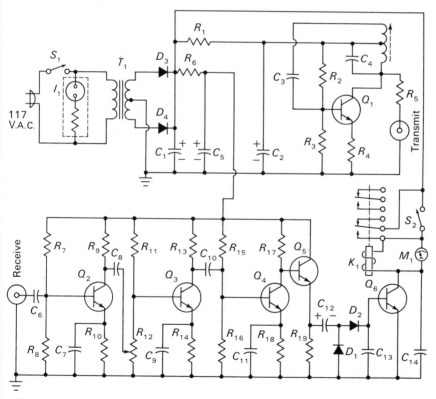

Figure 6.12
Circuit diagram for ultrasonic proximity detector.

C_1, C_5. 100-microfarad, 15-volt, electrolytic.

C_2. 30-microfarad, 15-volt, electrolytic.

C_3, C_7, C_9, C_{11}. 0.05-microfarad, 50-volt, ceramic disc.

C_4. 0.003-microfarad, 5% polystyrene.

C_6, C_8, C_{10}. 0.01-microfarad, 50-volt, ceramic disc.

C_{12}, C_{13}, C_{14}. 5-microfarad, 15-volt, electrolytic.

D_1, D_2. 1N34 germanium diode (or equivalent).

D_3, D_4. 50-volt PIV, 750-milliampere silicon rectifier.

I_1. Neon pilot light with built-in resistor.

K_1. Printed-circuit-type DPDT relay (Price Electric 206-14P, or equivalent).

L_1. 15-to-25-millihenry variable inductor with 10% tap (DEMCO 3E-027-1).

M_1. 0-to-15-volt D.C. voltmeter (Lafayette 99 G 5047, or equivalent).

Q_1, Q_6. 2N3706 (Texas Instruments, or equivalent).

Q_2, Q_3, Q_4, Q_5. 2N3708 (Texas Instruments, or equivalent).

R_1, R_4, R_6. 470-ohm, ½-watt.

R_2. 47,000-ohm, ½-watt.

R_3, R_{19}. 4,700-ohm, ½-watt.

R_5, R_{18}. 1,000-ohm, ½-watt.

R_7, R_{11}, R_{15}. 100,000-ohm, ½-watt.

R_8, R_{16}. 10,000-ohm, ½-watt.

R_9, R_{13}. 27,000-ohm, ½-watt.

R_{10}, R_{14}. 2,200-ohm, ½-watt.

R_{12}. 10,000-ohm trimmer (CTS X-201, or equivalent).

R_{17}. 15,000-ohm, ½-watt.

S_1, S_2. Miniature SPST toggle.

SO_1. 2-prong socket (optional).

T_1. Low-voltage transformer: 110-to-120-volt primary; 20-volt CT secondary (Stancor TP-2, or equivalent).

Chassis (Bud CU-465, or equivalent).

Circuit board (DEMCO #128).

Two 25-kilohertz ultrasonic transducers (DEMCO E-25).

Miscellaneous: Wire, solder, nuts, bolts, connectors, spacers, hardware.

Functional Description

The proximity detector consists of a transmitter that broadcasts an ultrasonic beam of sound and a receiver that detects the sound. The operating frequency is 25 kilohertz. Two small transducers, one from the transmitter and one from the receiver, are located to face each other or to reflect against a common surface and focus on each other. Any interruption of the beam of sound is detected by the receiver. The transmitter oscillator circuit consists of transistor Q_1 and its associated components. The receiver is a sensitive, four-stage amplifier (Q_2, Q_3, Q_4, Q_5, and associated components). The output stage consists of a half-wave voltage doubler (D_1, D_2, C_{12}, C_{13}) and transistor Q_6, which activates the control relay K_1.

Construction Notes

An actual-size photo of the printed-circuit board used for this project is shown in Figure 6.13. Parts layout on the printed-circuit board is shown in

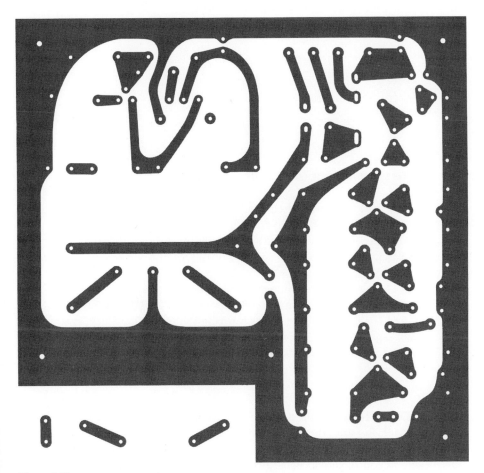

Figure 6.13
Actual-size photo of conductor side of printed circuit board (courtesy of Popular Electronics).

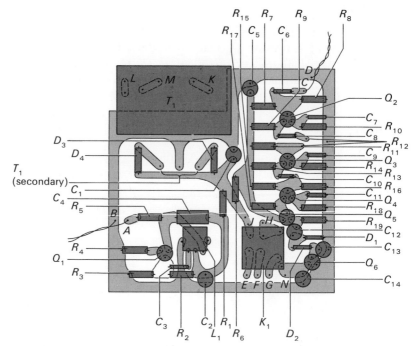

Figure 6.14
Parts placement on circuit board.

Figure 6.14. If a printed circuit is not used, avoid stray coupling between the transmitter and receiver sections, by treating the circuit as three independent circuits—power supply, transmitter, and receiver. Each may be assembled on its own chassis and interwired when mounted on the common chassis.

All associated wires routed from the circuit should be twisted together to avoid interaction, that is, leads from points A and B, C and D, and E, F, and G should be twisted when wiring the transmitter jack, the receiver jack, and the power switch, respectively. Operating relay K_1 can be located remotely, with twisted wires leading to its coil.

Operation Notes

Locate the transducers in the area to be monitored. Turn the sensitivity control, R_{12}, until the meter reads about 10 volts. The relay should activate. If it does not activate or if the meter reading does not reach 10 volts, the slug in coil L_1 should be adjusted, using a nonmetallic alignment tool. Set the slug about half-way into the coil form and then slowly turn it out of the form. As it is turned the meter reading should rise. If the reading reaches

10 volts, reduce the sensitivity and adjust the coil for a maximum voltage reading.

Caution

The transmitting transducers may create an audible noise when the circuit is fully adjusted. To eliminate this noise, carefully open the case of the transducer and insert a small piece of foam plastic under the crystal element, then close the case and seal it with rubber cement.*

---

* For the original description of this circuit, along with details on how to obtain the parts for it in kit form, see "Build the Ultrasonic Omni-Alarm," *Popular Electronics*, 24 (Apr. 1966), 41–45. Data sheets describing 40-kilohertz ultrasonic transducers are available from Massa Division of Cohu Electronics, 5 Fottler Road, Hingham, Mass. DEMCO parts are available from DEMCO, Box 16297, San Antonio, Texas 78216.

MAGNETIC PROXIMITY DRY-REED SWITCHES

Specifications

1. Contact ratings: available in types varying from 4 to 100 volt-amperes, with up to 500 volts breakdown at 60 cycles r.m.s.
2. Maximum switching speeds: approximately 925 hertz resonant frequency.
3. Available in normally open, normally closed, and double-throw configurations.
4. Operating life exceeds 15 million activations.
5. Switch may be activated by external permanent magnet or by electromagnetic field.
6. Operating temperatures: between about −65 and 150°C.

Applications

1. The switch can be used for rotational counting, as a response indicator, as a pressure indicator, and as a relay switch (see Figures 6.15–6.18).

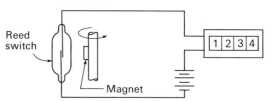

Figure 6.15
Rotational counting.

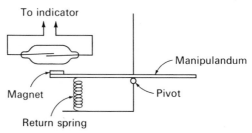

Figure 6.16
Response indicator.

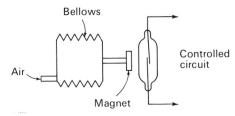

Figure 6.17
Pressure indicator.

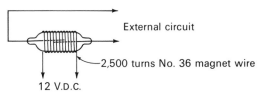

Figure 6.18
Relay switch.

2. The reed switch may be operated in the normally closed position by applying a magnetic field to the switch and then opposing this field with a permanent magnet (or electromagnet). The switch will open whenever the permanent magnet is placed near it and will close when the permanent magnet is removed.

Further Information

A large variety of reed switches is manufactured for special purposes other than those possible with the standard dry-reed switch. Reed relays that incorporate one or more reed switches in an operating coil are manufactured. There is a class of special reed relays employing multiple-wound coils that perform all the basic logic functions, including gating, flip-flop, and latching. Among special-purpose reed switches are the following.

1. Micro dry-reed switch. Rated at approximately 10 volt-amperes resistive (200-volt maximum). This is a smaller version of the standard dry-reed switch, for use where space is limited and power requirements are small. These relays are approximately 1½ inches long and 1/10 inch in diameter.

2. Mercury-wetted reed switch. Used to eliminate contact bounce. It is typically rated at approximately 50 volt-amperes resistive (400-volt maximum). This switch is about the same size as the standard reed switch of the same capacity. Since the switch contains a pool of mercury, which

flows from the bottom of the switch to the top by capillary action, it is position-sensitive.

3. High-voltage reed switch. Available with stand-off ratings up to 1,500 volts r.m.s. These switches are pressurized units with the same dimensions and contact ratings as the standard reed switch.

SCHMITT TRIGGER

Applications and Specifications

1. Sensitive voltage-level detector: A relay will close when the input voltage exceeds 2 volts and will open when the voltage falls below 1 volt.

2. Typically used to produce a square wave from a sinusoidal input.

3. Used to shape irregular waveforms into usable regular pulses in order to drive additional circuits.

4. Operating frequency: Approximately 20 hertz.

5. Power requirements: 20 to 28 volts D.C.

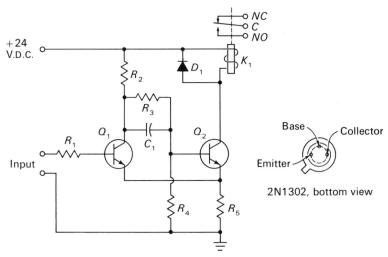

Figure 6.19
Circuit diagram of Schmitt trigger.

The circuit diagram is given in Figure 6.19. The parts list is as follows.

R_1. 1,000-ohm, ½-watt.
R_2. 5,100-ohm, ½-watt.
R_3, R_4. 3,300-ohm, ½-watt.
R_5. 15-ohm, ½-watt.
C_1. 0.01-microfarad, disc.
Q_1, Q_2. 2N1302.
D_1. 1N2069.

K_1. 24-volt D.C. relay; 400-ohm coil resistance (or greater).
Miscellaneous: Circuit foundation, power source, terminals, case, hardware.

Functional Description

With the biasing arrangement shown and with no input signal, transistor Q_1 is normally off, and transistor Q_2 is normally on, activating relay K_1. When a positive input signal greater than 2 volts is applied across R_1, Q_1 becomes biased on and Q_2 is biased off by R_2. When Q_2 is off, the relay current falls to zero and the relay opens. The circuit will return to its original state whenever the input voltage falls below the tripping level (approximately 1 volt) and the relay will again pull in.

Operation Notes

The 2-volt applied voltage is known as the upper trip point (UTP) and the 1-volt release voltage is known as the lower trip point (LTP). Both the UTP and LPT may be varied by changing the values of resistors R_3 and R_5.

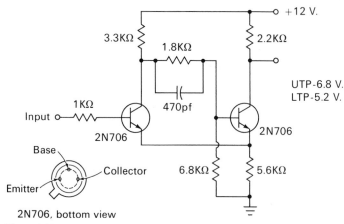

Figure 6.20
Typical high-frequency Schmitt trigger circuit.

The Schmitt trigger may be used in high-frequency applications (up to 1 megahertz) by replacing the control relay with higher-speed switching elements (transistors, control rectifiers, etc.). A typical Schmitt trigger to be used with a frequency of up to 1 megahertz is presented in Figure 6.20.*

---

* Precise design criteria for determining operating parameters as well as a detailed discussion of the design of Schmitt trigger circuits can be found in Texas Instruments, Inc., *Transistor Circuit Design* (New York: McGraw-Hill, 1963), pp. 381–83.

PULSE FORMER, MONOSTABLE MULTIVIBRATOR

Applications and Specifications

1. Pulse formers are used to convert brief signals, of any varying duration, from a few milliseconds upward, into uniform pulse signals of a prescribed duration.

2. Pulse formers are required when the response of an experimental subject is the activating source for programming circuits that employ digital logic modules that need uniform input signals.

3. Power requirements: 12 volts D.C. at 15 milliamperes.

4. Input: Positive-going pulse, from about 0.01 milliseconds to any length.

5. Output: Uniform relay-pulse signal of adjustable duration.

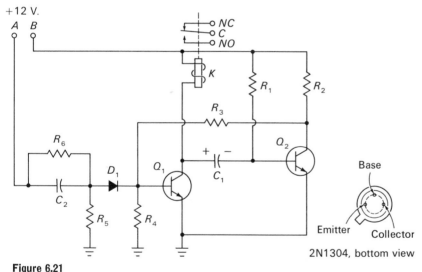

Figure 6.21
Circuit diagram of pulse former, monostable multivibrator.

The circuit diagram is given in Figure 6.21. The parts list is as follows.

R_1. See construction notes.
R_2, R_5. 1,000-ohm.
R_3. 10,000-ohm.

R_4. 47,000-ohm.
R_6. 2.2-megohm.
C_1. See construction notes.
C_2. 0.01-microfarad.
D_1. 1N2069.
Q_1, Q_2. 2N1304.
K_1. Sigma 4F1000 relay (or equivalent).
Miscellaneous: Circuit foundation, terminals, wire, hardware.

Functional Description

The monostable multivibrator requires an external signal voltage to operate. In its normal (no-signal) state, transistor Q_2 is held "on" by the base bias of R_1. When a positive trigger pulse is applied to the base of Q_1 through diode D_1, Q_1 gains sufficient forward bias to conduct. Capacitor C_1 begins to discharge through R_1 and in so doing reverse-biases the base of Q_2, preventing it from conducting. While Q_1 is conducting, relay K_1, which is the Q_1 collector load, is activated. When C_1 has discharged sufficiently, it can no longer maintain a reverse bias on the base of Q_2. Transistor Q_2 begins to conduct again and in so doing reverse-biases Q_1 through R_3. With Q_1 reverse-biased, relay K_1 is again shut off, completing the entire operating cycle.

The trigger circuit, consisting of R_6, C_2, R_5, and D_1, assures a sharp pulse at the base of Q_1 independent of the duration of the trigger input at points A and B, which may be connected to an external relay, switch, or other manipulanda that requires shorting to produce a trigger pulse. The time-delay period is governed by the values of C_1 and R_1.*

Construction Notes

The pulse duration may be approximated by the formula *pulse duration* $= 0.7\ RC$, where $R = R_1$ and $C = C_1$. With $R_1 = 50,000$ ohms and $C_1 = 1$ microfarad, the pulse duration is approximately 40 milliseconds.

A simple method of constructing the pulse-former circuit is illustrated in Figure 3.7.

If the pulse former is to be operated from negative-going pulses, the following modifications should be made: (a) substitute PNP transistors (2N1302) for the NPN transistors specified in the parts list; (b) reverse diode D_1 and capacitor C_1; (c) reverse the power-supply polarity.

---

* A pulse-former circuit using only relay circuitry is described in Appendix E.

SCR SENSOR CONTROL CIRCUITS

Applications and Specifications

1. Four switching circuits are presented for use with voltage sensors, resistance sensors, and switching devices. Type A is a simple D.C. latching switch for use with voltage sensors. Type B is a simple D.C. latching switch with capacitor turn-off, for use with voltage sensors. Type C is an A.C. static switch for use with A.C., D.C., or pulse triggers. Type D is an A.C. static switch for use with resistance or switch-type sensors.

2. Input characteristics of types A, B, C. Approximately 0.5 volts at 40 microamperes (20 microwatts) positive input signal will switch the SCR to a conducting mode. Input signal duration must be 1 microsecond or longer. Type C will operate from direct current, alternating current, or pulse signals.

3. Input characteristics of type D. Resistance sensor or switch closure controls the A.C. power delivered to a load. Type D is a solid state version of an A.C. relay; it eliminates the contact sticking, bounce, and wear that are common with relays.

4. Operating characteristics: type A, 200 volts D.C. at 1.5-ampere maximum; type B, 50 volts D.C. at 1.5-ampere maximum; type C, 200 volts A.C. at 1.5-ampere maximum; type D, 200 volts A.C. at 10-ampere maximum.

The circuit diagrams are given in Figures 6.22–6.25. The parts lists are as follows.

Type A

R_1. 1,000-ohm, ½-watt.
SCR. 2N2326.
Miscellaneous: Circuit foundation, terminals, wire, hardware.

Type B

R_1. 1,000-ohm, ½-watt.
R_2. 10,000-ohm, ½-watt.
C_1. Computed from the formula, $C_1 = 50 \times 10^{-6}/R_L$ (ohms).
SCR. 2N2323.
Miscellaneous: Contactor device, circuit foundation, terminals, wire, hardware.

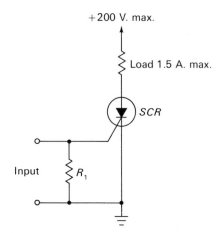

Figure 6.22
Circuit diagram for type A
SCR-sensor control circuit.

Type C

R_1. 1,000-ohm, ½-watt.
R_2. 47,000-ohm, ½-watt.
D_1. 1N1763.
SCR. 2N2326.
Miscellaneous: Circuit foundation, terminals, wire, hardware.

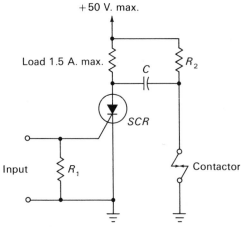

Figure 6.23
Circuit diagram for type B
SCR-sensor control circuit.

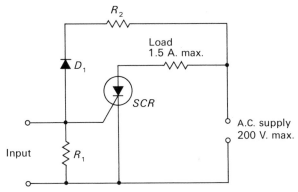

Figure 6.24
Circuit diagram for type C SCR-sensor control circuit.

Type D

R_1. 100-ohm, ½-watt.
R_2, R_3. 47-ohm, ½-watt.
D_1, D_2. 1N1692.
SCR_1, SCR_2. GE C36B.
Miscellaneous: Heat sinks for high-power use, circuit foundation,
   terminals, wire, hardware.

Functional Descriptions

Type A. The circuit, operating from D.C. power, functions as a simple
latching switch. The SCR, once triggered "on" through its gate lead, will
remain "on" until anode current is reduced below the dropout level
(effectively, zero). Resistor R_1 is a gate bias resistor, providing a negative
gate current that insures a stable "off" condition.

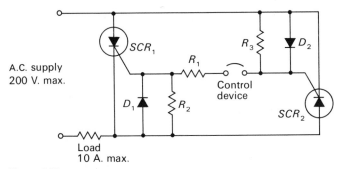

Figure 6.25
Circuit diagram for type D SCR-sensor control circuit
(courtesy of General Electric Company).

Type B. The SCR is triggered "on" as in the type A circuit. With the SCR conducting, capacitor C_1 charges through resistor R_2 to approximately the supply voltage. With the gate signal removed, the SCR may be shut off by closing the switch contacts, which causes the charge across the capacitor to drive the SCR anode negative with respect to ground, effectively turning "off" the SCR.

Type C. This circuit is designed to operate from A.C. supply voltages and to be triggered by either D.C. or A.C. gate voltages. Diode D_1 and resistor R_2 prevent excess heat from destroying the SCR. During the negative half-cycle of the supply voltage, gate voltages must be kept low because excess leakage currents might lead to thermal runaway.

Type D. In this circuit two SCRs are connected in parallel and in reverse configuration to each other. They are therefore capable of conducting during both half-cycles of the supply voltage, and applying nearly all the line voltage to the load. The gates are triggered by either shorting or placing a resistance across the control device contacts. The value of the resistance may be determined experimentally (with a 117-volt supply, the total resistance of R_1 and the control device should not fall below 100 ohms). Resistors R_2 and R_3 and diodes D_1 and D_2 limit the gate current and prevent inverse voltages from being applied between the cathode and gate during the reverse part of the cycle.

Construction Notes

When used with inductive loads (relays, solenoids, etc.), a diode should be placed across the load to suppress transient voltages that might otherwise damage the SCR. With the A.C. supply voltages used to power the type C and type D circuits, the transient voltage suppressor should be composed of two diodes (type 1694) placed back-to-back across the load.*

---

* For a detailed discussion of SCR devices, design parameters, and typical circuits, see General Electric Company, Semiconductor Products Department, *Controlled Rectifire Manual* (Syracuse, New York: 1960).

7

Control Circuits

Control circuits may be used to regulate the various operating steps in an experiment, such as the processing of information from sensing circuits and the subsequent activation of recording circuits. Control circuits are commonly used in behavioral experiments for regulating and controlling the stimulus and reinforcement contingencies. The control devices described in this chapter are conveniently divided into four categories: control amplifiers, regulating circuits, remote-control circuits, and control oscillators. Timing circuits and audio circuits, which are also commonly used for control purposes, will be discussed in separate chapters because of their special importance to the experimenter.

CONTROL AMPLIFIERS

Two types of amplifier circuits are described in this chapter. The first is a sensitive control amplifier capable of activating a relay in response to changes in input resistance or voltage. The second, a basic differential amplifier, is a useful null indicator for Wheatstone bridges and other D.C. sensing circuits.

The relay control amplifier is a two-transistor, direct-coupled amplifier that is wired to a sensitive relay. The simplicity of the circuit, together with its great reliability, permits incorporation of it directly into a large variety of circuits, some of which are shown in the circuit diagrams accompanying the amplifier description. A brief explanation of the amplification process used in control amplifiers is included in Chapter 1.

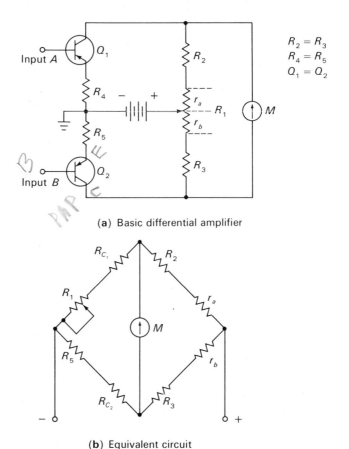

(a) Basic differential amplifier

(b) Equivalent circuit

Figure 7.1
Circuit diagram of (a) basic differential
amplifier and (b) an equivalent circuit.

The basic differential-amplifier circuit consists of two transistors wired in balance, as shown in Figure 7.1, a. Depending on the input level and polarity of the applied voltage, one transistor will conduct more than the other. Relative changes in conduction may be monitored by a meter (or a relay control amplifier stage) placed between the two transistors' collector leads.

The operation of the basic differential amplifier as a null indicator may be understood by referring to the equivalent circuit, Figure 7.1, b. In this conventional bridge circuit, the fixed resistances represent the load resistors, R_2 and R_3, and the two resistances of the balance potentiometer, r_a and r_b. The remaining two legs of the bridge represent the emitter resistors, R_4 and R_5, and the equivalent D.C. collector resistances of the transistors, R_{C_1} and R_{C_2}. Initially, the bridge is balanced by setting R_1 to show a zero meter reading, in order to compensate for any differences in the conducting characteristics of the transistors, as well as for small differences in the various resistors. A signal applied between either input A and ground, input B and ground, or input A and input B will change the equivalent D.C. collector resistances, R_{C_1} or R_{C_2}. These differences will unbalance the bridge, and result in a meter reading. As the input voltage or the voltage difference between two inputs approaches zero, the meter reading approaches zero, until the differences are nulled to zero.

Although the basic differential amplifier functions well as a null indicator in low-voltage direct-current applications, it cannot provide precise differential outputs during long periods of time. To provide such outputs, it would need precisely matched transistors, as well as a constant current source, which would tend to make the basic circuit complex and costly. Differential amplifiers are, nevertheless, extremely useful tools for the circuit designer. With the availability of integrated circuits, entire differential-amplifier circuits may be inexpensively constructed in packages no larger than a single standard transistor. The packaging of the entire circuit in a single case enables temperatures to be held virtually constant throughout the circuit, all but eliminates the changes in circuit performance that are caused by differences in temperature between circuit components, and results in extremely reliable and efficient amplifiers. Also, since the output of differential amplifiers is proportional to the differences between two input signals, the supply-voltage hum that produces ripple would affect both circuits equally and would not be reflected in the output, making the IC differential amplifier extremely useful in low-level applications, such as in EEG amplifiers.

The actual construction of IC differential amplifiers is not discussed in this book because it is better to design circuits for specific applications. However, several circuits using modified IC differential amplifiers for particular logic-gating, oscillator, and audio-amplifier applications are presented in Chapters 8 and 10. The reader should study the circuit descriptions to better appreciate the function and versatility of IC differential amplifiers.

REGULATOR CIRCUITS

Familiar regulator circuits include the thermostat or temperature regulator and the motor speed control or governor. These circuits are commonly used to maintain predetermined operating conditions for long periods of time and under various operating contingencies. Regulation is usually accomplished by using feedback from a sensing element to provide error information to a control element. These elements or components combine to form the regulating circuit.

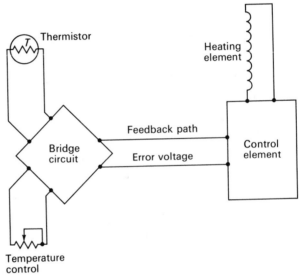

Figure 7.2
Block diagram showing functional elements
of a typical electronic thermostat.

Figure 7.2 is a block diagram of a common electronic thermostat, described in detail later in this chapter. A thermistor transducer and an adjustable temperature control are combined with two fixed resistors to form a bridge. A difference in resistance (an analog of temperature) between the value initially set by the adjustable control and the resistance of the thermistor produces an error voltage, which activates an electronic switching circuit (control element), which in turn activates a heater (or cooler), producing a temperature change that reduces to zero the resistance difference between the thermistor and the control. When there is no difference in resistance, the error voltage becomes zero and the control element becomes inoperative.

A variety of sensors may be used in regulating circuits. The distinction between regulating control circuits that use sensors and the sensing circuits described in Chapter 6 is, for the purposes of this book, that the control circuitry can directly handle load currents, whereas sensing circuitry must be connected to control circuits in order to control currents.

REMOTE-CONTROL DEVICE

The remote-control device described in this chapter permits the activation of a control relay that is not directly wired to the activator. In this application the 117-volt A.C. house wiring is used to transmit a high-frequency control signal to the receiver. The signal is capacitor-coupled to the house wiring and transmitted throughout it. A receiver plugged into a wall outlet detects the transmitted signal. The signal is rectified and applied through an amplifier stage to a control relay.

CONTROL OSCILLATORS

Uses of oscillator control circuits include: the generation of clock pulses as timing marks for programming circuits; as sawtooth wave oscillators in timing circuits and testing operations; and as pulse generators for stimulators. All the oscillator circuits described in this chapter consist of a reactive circuit element and an active circuit element. Capacitors are used as the reactive element and relays or semiconductors as the active element.

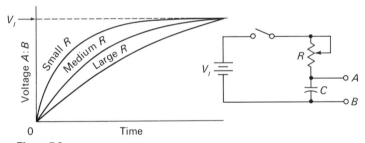

Figure 7.3
Rate at which a capacitor charges with various resistances.

The operation of the reactive element is illustrated in Figure 7.3. A capacitor, C, is charged exponentially through a resistor, R. For given values of applied voltage and capacitance, the time it takes the capacitor to charge to the given voltage is directly proportional to the resistance. If these times are plotted for various voltages and resistances, a graph like that in Figure 7.3 will result.

The active circuit element determines the shape of the output wave and conducts the wave to the output of the circuit. In the simple oscillator

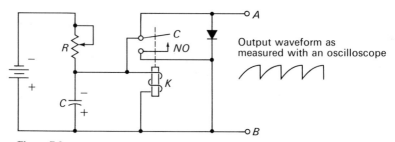

Figure 7.4
Sawtooth-wave relay oscillator.

shown in Figure 7.4, a relay is added to the output of the circuit shown in Figure 7.3, to be the active circuit element in a sawtooth wave relay oscillator circuit.

The capacitor, C, charges at a rate determined by R. At some point during the charging time, the voltage across C is enough to activate relay K. The contacts of relay K close when it is activated, and short-circuit the relay coil and capacitor, effectively discharging the capacitor and opening the relay. The output, as seen with an oscilloscope connected at points A and B, approximates a sawtooth wave. Using similar arrangements, as shown on p. 145, a relay, capacitor, and resistor may be used to generate square waves or pulses.

Solid-state devices may also be used as the active circuit element. For example, in a pulse generator described later in this chapter, a semiconductor known as a snap diode is used as the active circuit element to produce sharp rapid pulses. While the pulse generator's capacitor is charging, the snap diode remains nonconducting until the charge on the capacitor reaches the breakdown voltage of the diode, at which pooint the diode conducts and discharges the capacitor. The action of the snap diode is extremely rapid (in microseconds), both in turning on and in discharging the capacitor. A sharp pulse appears across a load connected in the diode circuit. When the capacitor is discharged, the snap diode becomes nonconducting and the capacitor again begins to charge. The charging rate, or operating frequency, is determined by the capacity of the capacitor and the resistance of the resistor used with the capacitor. Determining the actual values of resistance and capacitance used in various timing operations is discussed in Chapter 9.

ELECTRONIC RELAY CONTROL AMPLIFIER

Applications and Specifications

1. This versatile contact relay may be used as a variable-sensitivity drinkometer, photoelectric control relay, voltage sensor, time-delay relay, sound-actuated relay, liquid-level control, or other control source requiring extremely high sensitivity.

2. Sensitivity: Can be adjusted to activate with as much as 50 megohms across its resistance-sensitive input terminals, or with a control current of 0.2 microamperes at approximately 1 volt across its voltage-sensitive input terminals.

3. Power requirements: 6 to 9 volts D.C. at approximately 6 milliamperes with the relay energized.

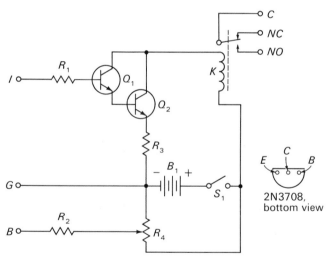

Figure 7.5
Circuit diagram of electronic relay control amplifier.

The circuit diagram is given in Figure 7.5. The parts list is as follows.

R_1. 1,000-ohm, ½-watt, 10%.
R_2. 10,000-ohm, ½-watt, 10%.
R_3. 47-ohm, ½-watt, 10%.
R_4. 1-megohm, 1-watt potentiometer.

Q_1, Q_2. 2N3708.
K. 1,000-ohm, sensitive relay (Sigma 4F1000, or equivalent).
S_1. SPST.
B. 9- to 12-volt battery.
Miscellaneous: Chassis, wire, hardware, terminals.

Functional Description

A two-stage, direct-coupled amplifier is used to drive a relay. Bias resistor R_3 stabilizes the circuit operation; R_4 and R_2 add bias and sensitivity control for operation with external resistive or switching sensors. Series resistor R_1 limits the base current to transistor Q_1 to prevent damage by strong input signals.

Construction Notes

Provide ample separation between the I, B, and G input terminals (about one-half inch between them), but parts layout is not otherwise critical. Use heat sinks on transistor leads when soldering. It is advisable to protect the circuitry by covering the circuit with either a plastic spray

Figure 7.6
Electronic relay control amplifier shown constructed
on an octal plug, forming a convenient plug-in module.

film (avoid spraying relay contacts) or a suitable dry enclosure to prevent erratic relay operation, which could be caused by shorting of the input terminals by moisture, metal filings, etc. Figure 7.6 shows the relay and associated circuitry mounted on an octal plug-in module.

Operation Notes

The basic wiring of the sensor to the electronic relay for sensing D.C. voltage outputs is hown in Figure 7.7; that for sensing high resistance is shown in Figure 7.8; and that for sensing low resistance is shown in Figure 7.9.

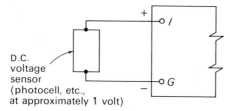

Figure 7.7
Sensing D.C. voltage outputs.

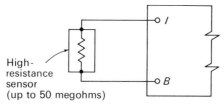

Figure 7.8
Sensing high resistance.

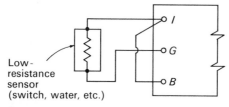

Figure 7.9
Sensing low resistance.

The wiring for a drinkometer setup is shown in Figure 7.10 (if the relay chatters, the I and B leads should be reversed; sensitivity may be adjusted by the sensitivity control), and that for a liquid-level control in Figure 7.11.

Wiring for time-delay relays is shown in Figure 7.12. In Figure 7.12, a, S_2 is a normally open, momentary-action, push-button switch. When S_2 is pressed, the relay is activated. When S_2 is released, the relay remains activated until C_1 is charged. With C_1 equal to 0.25 microfarad, and R_5 equal to 20 megohms, time delays between 3 and 9 seconds may be obtained

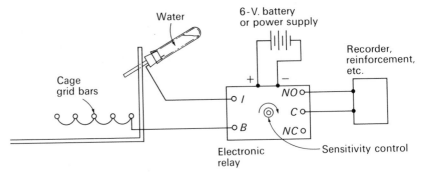

Figure 7.10
Drinkometer set-up.

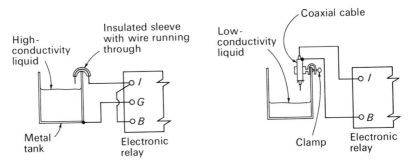

Figure 7.11
Liquid-level control.

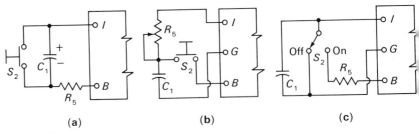

(a)   (b)   (c)

Figure 7.12
Time-delay relays.

by varying the setting of the sensitivity control. Capacitor C_1 may be discharged through the contacts of a second control relay wired in series with a 200-ohm surge resistor. In Figure 7.12, b, the capacitor is charged by the bias supply when S_2 is pressed and then discharges through R_5 and the relay input circuit when S_2 is released. The timing range is selected by choosing appropriate values for R_5 and C_1. In Figure 7.12, c, when S_2 is thrown to the on position, the relay will not be activated until C_1 is charged through R_5 sufficiently to reduce the voltage drop across R_5 and permit the relay to be activated. The timing range is selected by choosing appropriate values for R_5 and C_1.*

* A detailed description of this circuit, along with instructions for making a printed circuit and information for ordering a parts kit, can be found in "Super Sens," *Popular Electronics*, 23 (Nov. 1965), 57–61.

BASIC DIFFERENTIAL AMPLIFIER (NULL INDICATOR)

Applications and Specifications

1. A differential amplifier used to provide null indications in bridges and other sensing-circuit applications.

2. Provides a voltage gain of approximately 50 for accurate indications in low-output circuits.

3. May be modified to drive and control a relay amplifier stage.

4. Power requirements: 6 volts D.C., from a mercury battery or well-regulated power supply.

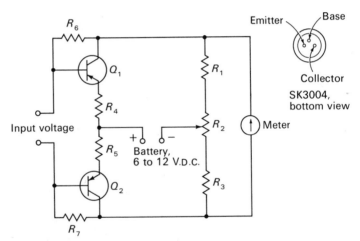

Figure 7.13
Circuit diagram of basic differential amplifier.

The circuit diagram is given in Figure 7.13. The parts list is as follows.

R_1, R_3. 2,200-ohm, $\frac{1}{2}$-watt, 5%.
R_2. 5,000-ohm, 2-watt potentiometer.
R_4, R_5. 220-ohm, $\frac{1}{2}$ watt, 5%.
R_6, R_7. 100,000-ohm, $\frac{1}{2}$-watt, 5%.
Q_1, Q_2. RCA-SK3004.
Meter. 100-0-100-microampere (center zero) meter movement (Allied type 52D7307 or equivalent).

Battery. Mercury battery or well-regulated power supply (Mallory type TR236R, 8.1-volt mercury battery, or equivalent).

Miscellaneous: Foundation, terminals, wire, hardware.

Functional Description

The amplifier circuit consists of two transistors, Q_1 and Q_2, wired with a common emitter and with bases to be fed voltages that are opposite in polarity. Depending on the polarity of the applied voltage, one transistor's collector current will increase and the other's will decrease. A meter placed across load resistors R_1, R_2, and R_3 indicates changes in applied voltage as well as in the polarity. Control R_1 is used to obtain an initial zero meter reading with no input signal. Resistors R_6 and R_7 provide base bias to the transistors, while resistors R_4 and R_5 tend to stabilize current flow through the transistors.

At the beginning of this chapter there is a brief description of differential amplifiers and some discussion of the limitations of their usefulness. The basic differential amplifier described here is not a precise laboratory instrument. Its primary function is as a null indicator in circuits that can be periodically adjusted for a zero reference voltage; such circuits are commonly used in voltmeters. It can be used with D.C. bridge circuits as well as A.C. null circuits.

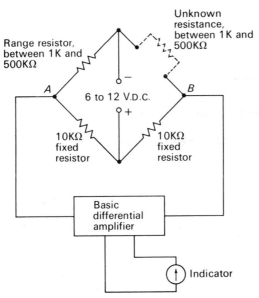

Figure 7.14
Simplified Wheatstone-bridge circuit connected to the basic differential amplifier.

Figure 7.14 is a simplified diagram of a typical Wheatstone-bridge circuit connected to the basic differential amplifier. As the unknown resistance varies, the currents in the bridge arms vary. The voltage difference between points *A* and *B* changes in direct proportion to the flow of current. The change in voltage is amplified and the output is used to drive an indicating meter or to activate a control stage.

Operation Notes

A control stage amplifier, such as the one described on p. 127, may be used in place of the meter to provide a control signal in response to D.C. input voltages.

The basic differential amplifier may also be used with certain A.C. null circuits. The circuit shown in Figure 7.15 may be used in place of the meter

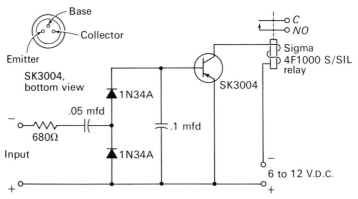

Figure 7.15
Relay-driver stage for basic differential amplifier.

to provide a control signal in response to A.C. input voltages. In this circuit, a voltage doubler is formed by the diodes, capacitors, and resistor. The doubler provides base bias to the transistor. For reliable operation, the driver stage should be operated from a separate or well-isolated power source. The values of the capacitors may require adjustment, depending on the operating frequency.

VARIABLE A.C. CONTROL

Applications and Specifications

1. Provides full control of incandescent lamps (up to 600 watts).

2. Provides full temperature control (up to 600 watts) of small heating elements, soldering irons, etc.

3. Provides speed control of small A.C. motors (shaded pole and universal motors, brush type) rated up to one-half horsepower.

4. Will not work effectively with inductive loads.

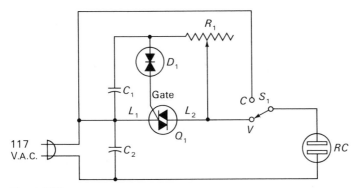

Figure 7.16
Circuit diagram of variable A.C. control. The C position of S_1 indicates constant voltage to load receptacle (RC); the V position indicates variable voltage (see text).

The circuit diagram is given in Figure 7.16. The parts list is as follows.

R_1. 250,000-ohm, 2-watt potentiometer.
C_1. 0.1-microfarad, 600-volt.
C_2. 0.05-microfarad, 600-volt.
D_1. GE ZJ-238 or Transitron ER-900 or Texas Instruments TI-43.
Q_1. GE ZJ-257 Triac.
Miscellaneous: Circuit foundation, heat sink, wall plug, receptacle, SPDT switch, wire, hardware.

Functional Description

With the function switch, S_1, set to "variable," capacitor C_1 is charged at a rate determined by the setting of control R_1. When C_1 charges to 30 volts, the pulse diode, D_1, fires and delivers a pulse to the gate terminal of the Triac, Q_1. This pulse turns the Triac on, permits full current to the load, and shorts out R_1 and C_1. The Triac will continue to conduct until the alternating-voltage cycle goes through zero, at which point the Triac will be cut off and C_1 can again begin to charge through R_1. Thus, by varying R_1, the firing angle of the Triac is adjusted and the effective load current is varied. Capacitor C_2 is added to prevent high-frequency Triac pulses from radiating down the power line and causing radio interference.

Construction Notes

A heat sink should be used with the Triac. An aluminum or copper piece at least $\frac{1}{8}$-inch thick and approximately $2\frac{1}{2}$ by 2 inches in size will form an adequate heat sink. If the heat sink is to be attached to the chassis, the Triac must be fastened to the heat sink in a manner that will insulate it electrically but permit thermal conduction (see Figure 7.17). Silicone

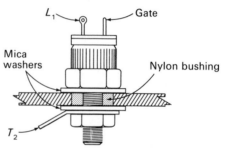

Figure 7.17
Mounting procedure for insulating the Triac from the heat-sink, while providing adequate thermal conduction.

grease should be spread over the mica washers, bushing, and Triac mounting stud to improve thermal conduction. Test electrical conductivity between Triac and heat sink with an ohmmeter to be certain there is no electrical connection.

Take care when soldering leads to Q_1 and D_1 to avoid overheating. Do not exceed the 600-watt rating of this circuit.

TEMPERATURE REGULATOR

Applications and Specifications

1. This circuit will control the amount of heat delivered by a heating element and will regulate the temperature of an environment by sensing changes in temperature. When the temperature rises above a predetermined point, the regulator will remove the heater from the power line; when the temperature falls below the predetermined point, it will connect the heating element to the power line.

2. Temperature control range: −6 to 75° C., ±1°.

3. Load control: 600 watts.

4. Power requirements: 120 volts A.C.

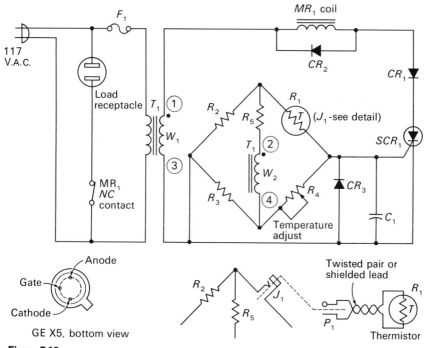

Figure 7.18
Circuit diagram of temperature regulator.

The circuit diagram is given in Figure 7.18. The parts list is as follows.

C_1. 0.05-microfarad, 200-volt.

CR_1, CR_2, CR_3. GE type 1N1693 rectifier diode.

F_1. 1-ampere fuse.

J_1. Temperature probe jack.

MR_1. Relay, DPDT, 5-ampere contacts with 6-volt D.C. coil (Potter & Brumfield GP11, or equivalent).

P_1. Temperature probe plug.

R_1. GE type 1D303 thermistor, 0.3-inch diameter, 1,000 ohms at approximately 25°C.

R_2, R_3. 1,000-ohm, 2-watt.

R_4. 2,500-ohm, 4-watt, wire-wound potentiometer.

R_5. 47-ohm, 2-watt.

SCR_1. GE-X5 silicon controlled rectifier.

T_1. Transformer: primary, 120 volts A.C.; secondary, W_1, 12.6 volts, and W_2, 12.6 volts (UTC-FT10, or equivalent).

Miscellaneous: Foundation, plug, wire, hardware.

Functional Description

Thermistor R_1 serves as one leg in a bridge circuit; the temperature adjustment control, R_4, serves as the second leg. When these two resistances are equal, no signal is available to the gate of SCR_1; relay MR_1 is not energized, and the load receives power through the normally closed contact of MR_1. When the temperature of the environment monitored by R_1 rises, the resistance of R_1 decreases, unbalancing the bridge and causing current to flow to the gate of SCR_1, turning SCR_1 on. Current through SCR_1 activates relay MR_1 and disconnects the load. When the temperature falls below the temperature that R_1 is set for, the bridge is unbalanced in the opposite direction, causing a negative signal to appear across the gate of SCR_1. The negative signal inhibits SCR_1 from firing, the relay's contacts close, and power is applied to the load. Transformer T_1 provides both relay control power and bridge power. The diodes and capacitor dampen transient voltages, and provide direct current for the SCR and the relay.

Construction Notes

The thermistor may be mounted in a plastic probe handle with only its sensing bead exposed to the environment. Use either twisted wire or shielded cable to connect the thermistor to the control box.

Operation Notes

The temperature adjustment control, R_4, may be calibrated by placing the thermistor in a water or oil bath with a reference thermometer. Heat the bath to some particular temperature, and rotate control R_4 until the relay drops out. Mark the control setting for that temperature. Calibrate for various other temperatures in the same manner, and place a permanent dial plate under the control pointer, calibrated in degrees.

The circuit may be modified to handle control relays with greater load-carrying capacity. Figure 7.19 shows an SCR being triggered through

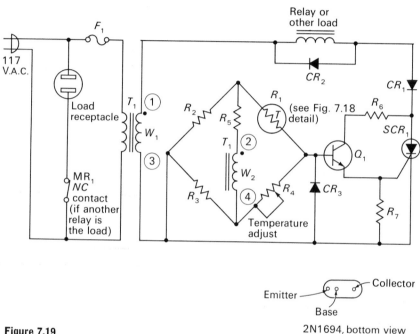

Figure 7.19
Circuit diagram of temperature regulator with added transistor stage and high-power SCR.

a transistor. The SCR shown will handle up to 4 amperes of current, enough to activate a higher-power control relay with higher-power contact ratings. Alternatively, the original circuit may be used with the relay acting as a pilot relay to activate a higher-power control relay. It is possible to substitute a cadmium sulfide light-sensitive photoconductor for the thermistor. In this mode, the control will turn on a light when the ambient light falls below a preset intensity.

Cautions

The SCR should be so located within the control box that it is not exposed to large changes in ambient temperature, which will affect its firing level. The leads of the SCR should be kept long, and it should not be situated within the box too close to any other components.*

---

* For a detailed description of this circuit, and a variety of practical, simple control circuits using SCRs, see General Electric Company, *Silicon-Controlled Rectifier Hobby Manual* (Auburn, New York, 1963), pp. 43–46.

REMOTE-CONTROL SWITCH

Applications and Specifications

1. Provides remote control of devices operating from 117-volt, 60-cycle currents.

2. Transmitter and receiver are coupled through the house wiring; radio-frequency signals activate a control relay at the receiver.

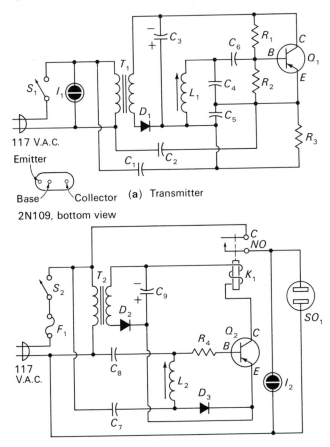

Figure 7.20
Circuit diagram of remote-control switch transmitter (a) and receiver (b).

3. Operating frequency: approximately 120 kilohertz.

4. Relay load current: approximately 150 milliamperes.

5. Power requirements: 117 volts.

6. This circuit should be usable wherever power-line outlets are available from the same fused circuit, and should operate reliably at distances up to 50 feet.

The circuit diagram is given in Figure 7.20. The parts list is as follows.

C_1, C_2. 0.005-microfarad, 1,000-volt, disc.

C_3. 20-microfarad, 50-volt, electrolytic.

C_4. 390-picofarad, mica.

C_5. 470-picofarad, mica.

C_6. 0.02-microfarad, disc.

C_7, C_8. 0.002-microfarad, 1,000-volt, disc.

C_9. 40-microfarad, 25-volt, electrolytic.

D_1, D_2. DD-117 (International Rectifier).

D_3. 1N48.

I_1. See S_1.

I_2. Indicator lamp (Allied Radio 8 U 698, or equivalent).

K_1. Miniature relay (Lafayette 99 C 6091, or equivalent).

L_1, L_2. 1.3 to 2.1-millihenry coil (Miller #4414).

Q_1, Q_2. 2N109.

R_1. 56,000-ohm, ½-watt.

R_2. 15,000-ohm, ½-watt.

R_3. 560-ohm, ½-watt.

R_4. 820-ohm, ½-watt.

S_1. SPST, with built-in indicator lamp (Lafayette 99 C 6259, or equivalent); or use separate switch and indicator lamp.

S_2. SPST.

SO_1, SO_2. Chassis-mounting receptacle.

T_1. 12-volt, 1-ampere, filament-type transformer.

T_2. 6.3-volt, 1-ampere, filament-type transformer.

Miscellaneous: Two bakelite boxes, four A.C. line cords, perforated board, terminals, spacers, wire, hardware.

Functional Description

The transmitter consists of a Colpitts oscillator whose frequency is determined by an adjustable coil, L_1, and capacitors C_4 and C_5. The signal is capacitor-coupled to the A.C. line by C_1 and C_2. Power for the oscillator circuit is provided by transformer T_1. The low-voltage output of T_1 is rectified by diode D_1 and filtered by capacitor C_3. Switch S_1 activates the entire circuit, and S_2 is the control switch for the receiver.

The receiver detects the signal from the transmitter through a tuned circuit, C_7, C_8, and L_2. This signal is rectified by D_3, and used to provide operating bias for transistor Q_2. When Q_2 conducts, relay K_1 energizes and connects the load directly to the power line through SO_1 and the *NO* contacts of the relay. Switch S_2 is used to disable the receiving circuit.

Construction Notes

Parts layout of neither the receiver nor the transmitter is critical. If metal boxes are used, take care to prevent short circuits to the metal case. No leads should be connected to a metal case. Provide access holes in line with the tuning slugs of coils L_1 and L_2 to facilitate tuning.

Operation Notes

Turn both the receiver and transmitter on. If relay K_1 does not immediately pull in, adjust L_1, L_2, or both until the relay energizes.

With a voltmeter placed across the relay coil, adjust L_1, L_2, or both for a peak voltage indication. With the receiver and transmitter close-coupled, the voltage should read between about 6 and 10 volts D.C.*

---

* For further information about this circuit, see "Build a Carrier-Current Remote-Control System," *Popular Electronics*, 26 (Jan. 1967), 50–53.

BASIC RELAY PULSE GENERATORS AND PULSE SHAPERS

Applications and Specifications

1. Provide uniform and repetitive pulses for counting, controlling, programming, or simulating animal responses.

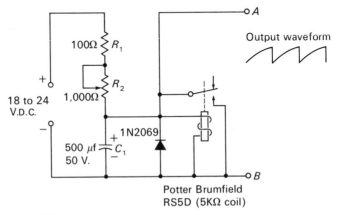

Figure 7.21
Sawtooth-wave generator.

2. The circuit of Figure 7.21 provides sawtooth waves with a frequency adjustable from approximately 1 to 20 hertz. This circuit is useful for providing timing or triggering pulses for controlled rectifier or transistor circuits, such as flip-flops and counters.

3. The circuit of Figure 7.22 provides square waves with a frequency adjustable from approximately 1 to 20 hertz. This circuit is useful in controlling auxillary relay or solenoid circuits, such as those associated with stimulus presentation devices, programming circuits, and relay counters.

4. The circuit of Figure 7.23 is a simple pulse generator with a frequency adjustable from approximately 1 to 20 hertz. This circuit is shown with its output directly coupled to a pilot lamp for use as a light flasher.

5. The circuit of Figure 7.24 is a pulse-shaping circuit. With the values shown, it can provide a uniform pulse of approximately 2 seconds duration when triggered by external pulses of shorter duration. It can be used to convert input pulses of varying duration into uniform output pulses. This circuit should not be confused with a pulse-forming circuit, such as that

described in Chapter 6. The pulse-forming circuit is useful when converting input pulses of varying duration, up to and including a constant input, into uniform output pulses. The circuit described here will not produce uniform output pulses when the input pulse is longer than the relay closing time, or approximately 5 milliseconds. In addition, the triggering pulse must be a voltage capable of activating the relay.

6. Power requirements: All circuits are shown with supply voltages between 18 and 24 volts D.C. Other voltages can be used, if suitable changes are made in the associated relay-coil circuits.

Functional Description

1. Sawtooth-wave generator (Figure 7.21). In this circuit, R_1 functions as a current-limiting resistor, while R_2 adjusts the rate of charge on capacitor C_1. As the capacitor is charging, the output across points A and B rises at the charging rate of the capacitor. When the voltage reaches the firing level of the relay, the normally open relay contacts close, effectively clamping the output to ground and discharging the capacitor. The value of the peak output voltage is determined by the value of R_1. Although the sawtooth wave produced at points A and B is not perfectly linear, it is suitable for many control purposes.

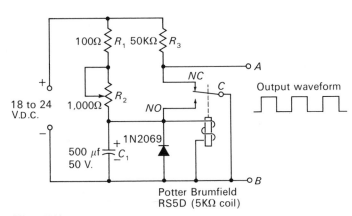

Figure 7.22
Square-wave generator.

2. Square-wave generator (Figure 7.22). In this circuit, R_1 and R_2 function as described for the sawtooth-wave generator. The output across A and B alternately samples the line voltage through resistor R_3 and is then shorted by the normally open relay contact. Resistor R_3 prevents the power supply from being short-circuited when the output is shorted through the normally closed contact, and aids in equalizing the pulse width.

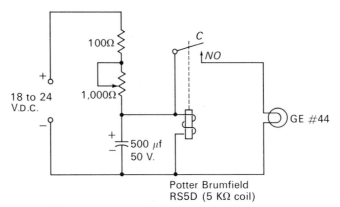

Figure 7.23
Pulse generator (light flasher).

3. Pulse generator (Figure 7.23). This circuit is a variation of the saw-tooth-wave generator, and its operation is identical except that the output is taken directly across the relay coil. The circuit can function as an effective metronome if the pilot lamp is replaced with a loudspeaker, which will produce an audible click each time the relay contacts closed.

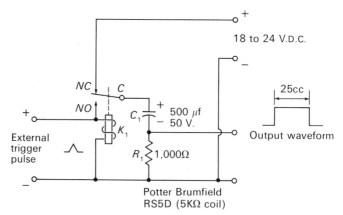

Figure 7.24
Pulse shaper.

4. Pulse generator and shaper (Figure 7.24). When a 24-volt pulse appears across relay K_1, the relay pulls in and connects capacitor C_1 and resistor R_1 across the relay coil. Relay K_1 will hold in until capacitor C_1 has discharged through the relay, R_1, and the load in parallel with R_1, at which point K_1 will release and be set for another trigger pulse. The output pulse is approximately 2 seconds. This duration may be changed by varying the value of C_1.

Further Information

In all these circuits, a sensitive relay containing SPDT contacts is employed. Other relays, such as sensitive telephone-type relays, are available with a variety of contact arrangements for use where additional control functions are desired. The coil resistance and voltage ratings of these relays may require changes in the values of the capacitors and resistors or in the supply voltage.

ASTABLE MULTIVIBRATOR OSCILLATOR

Applications and Specifications

1. High-efficiency square-wave generator for timing, counting, and response simulation.

2. Frequency is adjustable between about 50 hertz and 50 kilohertz.

3. Power requirements: 6 to 12 volts D.C. at 2 milliamperes.

4. Output: Square wave, with voltage variable from zero to about 2 volts, peak to peak. Output impedance is less than 1,000 ohms.

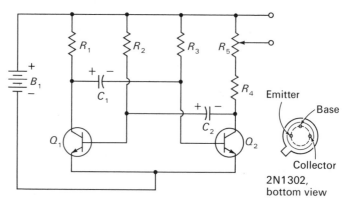

Figure 7.25
Circuit diagram of a stable multivibrator oscillator.

The circuit diagram is given in Figure 7.25. The parts list is as follows.

R_1, R_4. 4,700-ohm, ½-watt.
R_2, R_3. 150,000-ohm, ½-watt.
R_5. 1,000-ohm potentiometer.
C_1, C_2. See text.
Q_1, Q_2. 2N1302.
B_1. 6 to 12 volts D.C.
Miscellaneous: Foundation, terminals, wire, hardware.

Functional Description

The astable or free-running multivibrator does not require an external signal in order to operate. When power is applied to the circuit, one transistor conducts while the other is cut off. How long the transistor conducts is determined largely by the values of resistors R_2 and R_3 and capacitors C_1 and C_2. The functioning of the circuit may be understood by considering how the transistors shift in conduction. When power is applied to the circuit, one transistor begins to conduct sooner than the other transistor, because of slight irregularities in the symmetry of the circuit. Assume that collector current begins to flow in transistor Q_1. As the collector current increases, there is an increase in the voltage across resistor R_1, with the collector end of the resistor becoming more negative. Capacitor C_1, connected to the collector end of R_1, attempts to become more negative, and the base of transistor Q_2 becomes negatively biased. The net result is a decrease in current through transistor Q_2 and R_4.

With the voltage lowered across resistor R_4. capacitor C_2 starts to become less negative, that is, more positive. This apparent increase in positive potential is applied to the base of transistor Q_1, causing more collector current to flow in Q_1. This process continues until C_2 is charged and transistor Q_2 is cut off.

Transistor Q_2 remains cut off until the charge on C_2 begins to leak off. The discharge path of C_2 is through the base circuit of Q_2. When C_2 has discharged sufficiently, collector current begins to flow through R_4, causing the collector end of R_4 to become increasingly negative. The base of Q_1 is made increasingly negative, causing the current through R_1 to decrease. As the current through R_1 decreases, the corresponding voltage drop across R_1 causes the base of Q_2 (through C_1) to become positively biased. The increased current through R_4 and R_5 rapidly drives Q_1 to cut off (through C_2). When the charge on C_1 begins to leak off, the process begins again.

Construction Notes

The frequency of the oscillation is determined by the values of capacitors C_1 and C_2 and of resistors R_2 and R_3. The frequency may be approximated by the formula $f = 0.73/RC$, where f is the frequency in hertz, $R = R_2 = R_3$ in ohms, and $C = C_1 = C_2$ in farads. Table 7.1 lists sample values of

Table 7.1
Values of Capacitance for Various Frequencies.

Frequency in hertz	C_1, C_2
10,000	470-picofarad
1,000	0.005-microfarad
400	0.0125-microfarad
50	0.1-microfarad

C_1 and C_2 to be used for particular frequencies within the human audible range (R_2 and R_3 are both 150,000 ohms). The capacitor values listed in Table 7.1 are only approximations of the ones that will give the desired frequency. To establish the frequency precisely, the circuit could be modified by substituting a ganged set of potentiometers for R_2 and R_3. The output of the oscillator could then be calibrated against a laboratory standard by using an oscilloscope to observe the outputs of both oscillators, and adjusting the constructed oscillator frequency by means of the potentiometer.

This circuit is a symmetric multivibrator oscillator. Any one transistor's conducting or on time is equal to its nonconducting or off time; that is, it has a 50 per cent duty-cycle. It is possible to construct an oscillator with other duty cycles by varying the values of the various timing capacitors and resistors asymmetrically, in proportion to the duty cycle required. If, for example, one wanted to maintain the Q_2 transistor in its on state for one-fourth of the time it is in its off state, then the value of C_1 could be selected to be one-fifth that of C_2.

The prolonged operating accuracy of the multivibrator oscillator circuit depends on the stability of the power supply as well as on the quality of the capacitors employed.*

---

* For a detailed discussion of multivibrator design, see Texas Instruments, Incorporated, *Transistor Circuit Design* (New York: McGraw-Hill, 1963), pp. 369–83.

PULSE GENERATOR

Applications and Specifications

1. Provides variable-frequency trigger pulses for counting and triggering of associated equipment.
2. Pulse rate is adjustable with range capacitors, from 1 to 10,000 pulses per second; a potentiometer provides precise pulse frequencies within each range.
3. Output voltage: variable between zero and about 10 volts.
4. Rise time of pulse is in nanoseconds. Pulse width is approximately 0.001 times the pulse frequency.
5. Power requirements: 24-volt D.C. battery or well-regulated power supply.

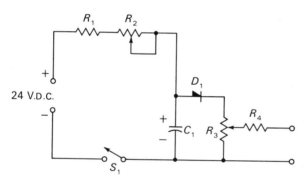

Figure 7.26
Circuit diagram of pulse generator.

The circuit diagram is given in Figure 7.26. The parts list is as follows.

D_1. Motorola M4L3054 (or equivalent).
R_1. 47,000-ohm, ½-watt.
R_2. 500,000-ohm, 2-watt potentiometer.
R_3. 500-ohm, 2-watt potentiometer.
R_4. 6-ohm, ½-watt.
S_1. SPST.
Miscellaneous: Foundation, terminals, wire, hardware.

Functional Description

A four-layer diode, D_1, is used in a conventional R-C timing circuit. The four-layer diode acts as a voltage-sensitive switch. When the charge on the timing capacitor reaches the diode's operating voltage, the diode rapidly discharges the capacitor through a voltage divider (R_3), a current-limiting resistance (R_4), and the output. Since resistances R_3 and R_4 are smaller than the combined timing resistances of R_1 and R_2, the discharge of the capacitor through the diode will cause current flow through the diode to fall rapidly below the firing level of the diode, which will turn "off" when the capacitor voltage is near zero. Approximately 1 microampere or more of current flow is required to maintain conduction through the diode. When the diode is turned "off," the capacitor will again begin to charge toward the supply voltage, repeating the operating cycle.

Operation Notes

Table 7.2 lists sample values of C_1 to be used for particular frequency

Table 7.2
Values of Capacitance for Various Pulses.

Frequency, in pulses per second	Capacitor C_1, in microfarads
10	3, 50-volt electrolytic
100	0.3, mylar
1000	0.03, mylar

ranges (the values are only approximations of the ones needed for precise frequencies). Control R_2 functions as a vernier frequency control within each frequency range.

8

Programming Circuits

Programming circuits determine the sequence of the steps in an automated experiment. The sequencing is accomplished by high-speed electronic switching circuits, which can respond to input signals in a predetermined, logical manner. Programming circuits may consist of: logic-gating circuits, which route signals discretely; oscillator circuits, used as clocks or timers; flip-flop circuits, used as memory devices; pulse-forming circuits, used to shape signals; and driver circuits, used to couple the programming circuits to other circuits.

By combining various programming circuits, high-speed computing devices may be constructed. These devices may be able to process enormous quantities of information within very short times. Computing machines are composed of simple gating and memory circuits, which are combined to perform various logic functions. The basic computing machine also needs input devices to provide operating signals, output devices for delivering information from the machine, and devices for storing operating instructions and data until they are needed.

Fundamental to both computing machinery and programming circuits are basic logic-gating circuits, which are discussed in detail in the following pages. Flip-flop circuits, driver circuits, and integrated-circuit programming devices are also included, as separate construction projects. Because of the importance of the pulse former in the initial control stage of sensing circuits, a description of a practical pulse-former circuit was included earlier, in Chapter 6. Also, because of the importance of the square-wave oscillator in certain control operations, a description of a practical square-wave oscillator was included in Chapter 7.

The combining of various logic circuits into computing machines, counting machines, or special-purpose programming devices is not within the scope of this book. The bibliography at the end of the book lists several sources that describe computing devices and logic circuits.

BASIC LOGIC-GATING CIRCUITS

The basic logic gate may consist of a simple manual switching circuit, of relay switching circuitry, of solid state switching circuits capable of extremely fast and accurate switching speeds, or of combinations of all three. Combining various gating circuits enables a system of predetermined logical functions to be performed in precise accordance with a programmed plan. Familiarity with a variety of gating circuits will enable the experimenter to program and implement an experimental plan rapidly, efficiently, and with a minimum of complexity and cost.

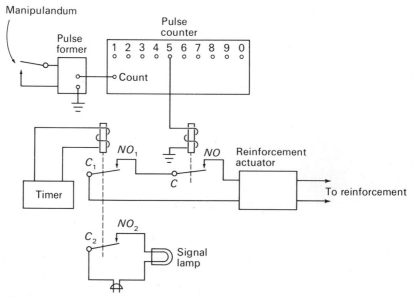

Figure 8.1
Reinforcement-schedule circuit using relay gating.

Figure 8.1 illustrates the use of a simple logic-gating circuit in a reinforcement schedule. Relay switching is used to dispense a reinforcement to the subject after the fifth press of a lever, but only if a predetermined time interval has elapsed, as marked by the turning on of a signal light. In this circuit two relays are used to carry out the program. The contacts of the relays are wired in series; both must be closed before the reinforcement activator circuit can deliver a signal. The pulse former is used to deliver a constant signal to the pulse counter. A second set of contacts on the relay connected to the timer activate the signal lamp at the end of the timing interval.

Another gating method could be used to carry out the above program. This method, shown in Figure 8.2, employs a multiinput, solid state *AND*

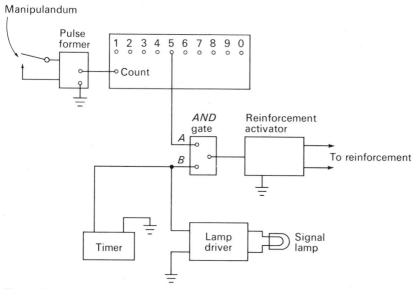

Figure 8.2
Reinforcement-schedule circuit using *AND* gate.

gate, which replaces the series-connected relay contacts of Figure 8.1. The *AND* gate will deliver a reinforcement signal only when an input signal is simultaneously present at both of the two input positions, *A* and *B*. Figure 8.2 also shows a lamp driver, used to activate the lamp when a signal comes from the timer. Lamp-driver circuits will be discussed later in this chapter.

The advantages of the solid state *AND* gate become obvious when many functions must be performed simultaneously before a reinforcement is presented. Additional inputs to the solid state *AND* gate can be easily constructed, and are much more reliable, much cheaper, and much less bulky than the relay circuitry that would be needed to perform the same function. Furthermore, one can create an inventory of solid state logic gating circuits to be used as building blocks in the creation of a variety of programs. These building blocks include *AND* gates, *OR* gates, *NAND* gates, and *NOR* gates. These gating circuits and their functions will be described in the following section, along with analogous gating circuits using switches and relays.

The basic gating circuits we will describe all operate on the binary principle; that is, they function in one of two modes, signal, or no signal. The signal mode consists of an electrical input to the logic gate that is of sufficient voltage and of the correct polarity to enable the gate components to conduct or not conduct a current to the output terminal, depending on the required function of the gate. The no-signal mode consists of no electrical input to the logic gate; this lack of input enables the gate components to conduct or not conduct a current to the output terminal, depending on the required function of the gate.

Circuit Descriptions

In the following circuit descriptions, the logic function performed by a particular gating circuit is explained by the use of a "truth table" (see Figure 8.3), which is a statement of all possible operational modes of the gate, based on whether or not a signal is present at the inputs to the gate. A signal is defined as binary 1; no signal is defined as binary 0. In addition to the truth table, Figure 8.3 gives a symbolic diagram of the particular gate whose logic is described in the truth table, and a logical equation (known as a Boolean statement, after the nineteenth-century logician, George Boole) that expresses the relationships between the inputs and outputs of the gating circuits, for each of the most common logic gates used in programming circuits. The Boolean statement for the *AND* gate may be read, "*C* is achieved only by both *A* and *B*." For the *OR* gate the Boolean statement may be read, "*C* is achieved by either *A* or *B* or both." For the *NOT* circuit the Boolean statement may be read, "*B* is achieved when *A* is not." The *NOT* gate is also known as an inverter gate. The *NAND* gate is represented symbolically by an *AND* gate in combination with a *NOT* gate. The Boolean statement for the *NAND* gate may be read, "*C* is achieved only by not *A* and not *B*." The outputs of the *NAND* truth table are opposite to those of the *AND* truth table. The *NOR* gate is represented symbolically by an *OR* gate in combination with a *NOT* gate. The Boolean statement for the *NOR* gate may be read, "*C* is achieved by either not *A* or by not *B* or both not *A* and not *B*." The outputs of the *NOR* truth table are opposite to those of the *OR* truth table. In each of the

Gate	Truth table	Symbolic diagram	Logical equation
AND	Input || Output A|B||C 0|0||0 0|1||0 1|0||0 1|1||1	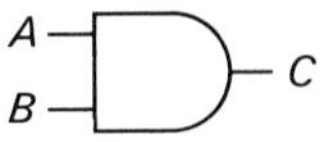 A, B → C	$C = A \cdot B$
OR	Input || Output A|B||C 0|0||0 0|1||1 1|0||1 1|1||1	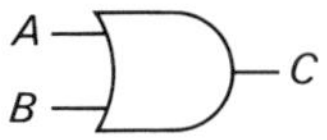 A, B → C	$C = A + B$
NOT	Input || Output A||B 0||1 1||0	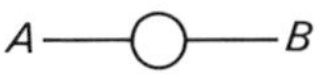 A → B	$B = \overline{A}$
NAND	Input || Output A|B||C 0|0||1 0|1||1 1|0||1 1|1||0	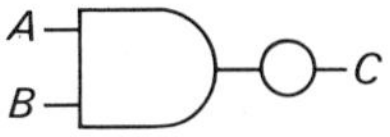 A, B → C	$C = \overline{A \cdot B}$
NOR	Input || Output A|B||C 0|0||1 0|1||0 1|0||0 1|1||0	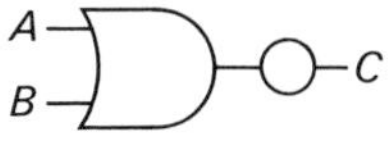 A, B → C	$C = \overline{A + B}$

Figure 8.3
Logic gates, truth tables, symbolic diagrams, and logical equations.

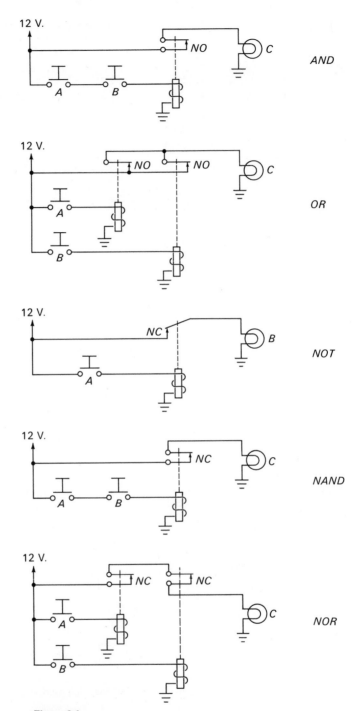

Figure 8.4
Relay gating circuits.

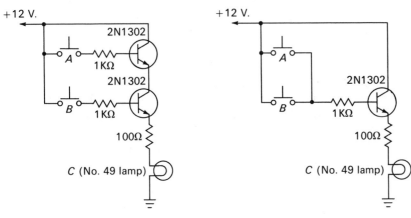

(a) Basic transistorized *AND* gate   (b) Basic transistorized *OR* gate

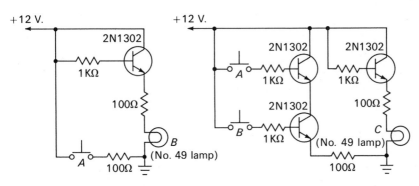

(c) Basic transistorized *NOT* gate   (d) Basic transistorized *NAND* gate

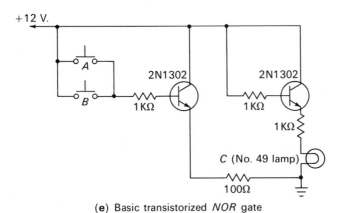

(e) Basic transistorized *NOR* gate

Figure 8.5
Basic transistorized gating circuits.

gates described, except the *NOT* gate, the number of inputs to any particular gate may be greater than two. The *NOT* gate inverts the state of any expression introduced to its input.

Figure 8.4 shows logic-gating circuits that use relays, switches, and lamps to accomplish the logic functions described in Figure 8.3. In each of the relay circuits, a normally open switch corresponds to a binary 0 input, a normally closed switch to a binary 1 input. The extinguished lamp corresponds to a binary 0 output, the lighted lamp to a binary 1 output. The function of each relay gate may be understood by considering the operation of the gate when inputs are applied corresponding to the possibilities listed in the truth table of Figure 8.3.

Figure 8.5, a, shows a basic transistorized *AND* gate. Before an output can turn on a signal light, both input switches, *A* and *B*, must be closed to provide base biasing through the 1,000-ohm resistors connected to each transistor. Without the base bias for each transistor, no output signal is possible, since the emitters and collectors of transistors are wired in series with the signal lamp.

Figure 8.5, b, shows a basic transistorized *OR* gate. When either input switch *A* or input switch *B* (or both) is closed, the transistor will be biased to conduct through the 1,000-ohm resistor, causing the output signal to appear. When both input switches are open, the transistor will not conduct enough to produce an output. Any number of inputs may be wired in parallel to inputs *A* and *B* for multiinput operation.

Figure 8.5, c, shows a basic transistorized *NOT* gate. With input switch *A* in the open position (binary 0), the transistor is biased to conduct through the 1,000-ohm resistor, providing an output signal light (binary 1). When input switch *A* is closed, the transistor base-to-emitter lead is essentially short-circuited, cutting off the transistor and removing current from the lamp. By combining the basic transistorized gates described above, *NAND* and *NOR* gates may be easily constructed, as shown in Figures 8.5, d, and 8.5, e.

Combining Circuits

In each of the basic solid state gating circuits described above, the input signal is provided by a switch. Often one wants the signal to appear in the form of an electrical pulse or a series of pulses. One may also want to couple together various gating circuits to perform a logic function at high switching speeds, as determined, for example, by a stable oscillator operating at high frequencies. In a digital computer, complex arithmetic equations are solved by gating circuits that perform simple mathematical functions at very high switching speeds. Results of the simple mathematical functions are either stored in a memory for later processing, or are routed through other gating circuits to be combined with other mathematical functions, as determined by the computer program. Digital-computer circuitry is designed to operate with discrete signals of specific voltages and polarity. Similarly, when using logic-gating circuits for experimental designs, one wants all the circuitry to operate at specific electrical values, in order to simplify the design criteria for power sources and associated equipment. Examples of gating circuits designed for high-speed switching operations are presented below.

The following circuits represent gating circuitry that can perform the logical *AND, NAND, OR,* or *NOR* functions given discrete inputs in the form of high-speed switching signal voltages. These circuits differ from the basic logic gates described above in the manner in which signals are conducted through the gate. The following circuitry is known as resistor-transistor logic or *NOR* logic. The basic circuit used to create the various logic gates is the *NOR* gate, shown in its basic form in Figure 8.6. A

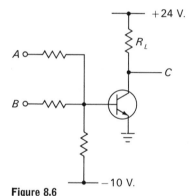

Figure 8.6
Basic *NOR* resistor-transistor
logic gate.

second source of voltage (of minus 10 volts) is used in the transistor base circuit. This voltage is used to insure that the transistor is held in its cut-off state until the input voltage is high enough to turn the transistor on and cause it to conduct. The second source of voltage also serves the important function of stabilizing the transistor's characteristics under various operating conditions and temperatures.

In the following discussion, an input signal is defined as binary 1 when a positive voltage of 24 volts is applied to an input terminal. An output signal is defined as binary 1 when a positive voltage appears at point *C*; it appears only when the transistor is cut off. Conversely, when the transistor is biased into conduction by the application of a positive base-bias voltage, a short circuit is effectively formed between the collector and the

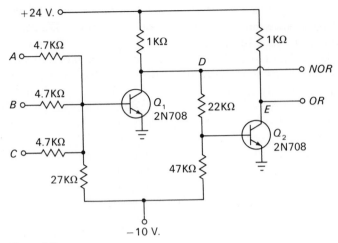

Figure 8.7
NOR, OR logic-gating circuit.

emitter. Thus when the input is positive, at *A* or *B* or both, the output is near ground potential, corresponding to binary 0. The gate thus performs the logical *NOR* function described in the truth table (Figure 8.3) since either or both of the input signals will result in a binary 0 (grounded output), while no signal input results in a binary 1 (positive output voltage).

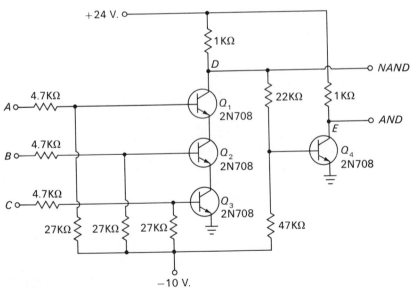

Figure 8.8
NAND, AND logic-gating circuit.

A practical resistor-transistor logic-gating circuit is shown in Figure 8.7. It can be seen that the basic *NOR* signal gate is used to perform the *NOR* function if the signal is taken from point *D*. Figure 8.7 also shows an additional stage that can invert the input signal. This inverting stage functions similarly to the basic *NOT* gate described earlier. By inverting the basic *NOR* function, the logical *OR* function results. Either function of the gating circuit can be chosen for a given operation.

In Figure 8.8, the basic *NOR* resistor-transistor logic gate is used to fulfill the logical *NAND* and *AND* functions. Each input goes to the base of a separate transistor. The transistors in turn are series-connected through their emitters and collectors. With no input signal present at any input, a positive voltage is present at point *D* (binary 1). With 24-volt input signals at some but not all inputs, a positive voltage will be present at point *D* (binary 1). When 24-volt input signals are applied simultaneously to *A* and *B* and *C*, the transistors conduct and point *D* is near ground potential (binary 0). These conditions fulfill the criteria for the *NAND* gate as defined by the truth table of Figure 8.3. By adding an inverter stage (Q_4 and its associated resistors), the logical *AND* function results. Again, either function of the gating circuit can be chosen for a given operation.

The resistor-transistor logic is an attractive way of creating logic-gating

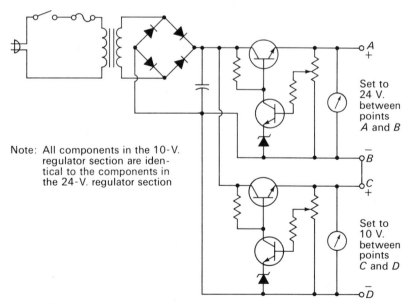

Note: All components in the 10-V. regulator section are identical to the components in the 24-V. regulator section

Set to 24 V. between points *A* and *B*

Set to 10 V. between points *C* and *D*

Figure 8.9
Modification of power supply (pp. 66–67) to provide dual voltage for *NOR* logic gates.

circuits, since it uses only a few, inexpensive components to create the various logic functions. Other methods of high-speed gating include diode logic (using diodes to route signals discretely), direct-coupled transistor logic (similar to the basic logic circuitry described above), and various combinations of diodes and transistor circuitry. Integrated circuits are also used extensively in computer circuitry, since they occupy little space, consume little power, and are extremely fast-acting and stable. Examples of integrated-circuit logic circuits are presented later in this chapter.

A stable power supply suitable for powering the circuits described here can be found on p. 66. Figure 8.9 illustrates a simple method of modifying it to provide the 10-volt bias voltage required for the resistor-transistor logic circuits, as well as the 24-volt operating voltage.

BISTABLE MULTIVIBRATOR (FLIP-FLOP)

Applications and Specifications

1. The flip-flop circuit forms the basic design element in binary counters, shift registers, and memory circuits.
2. Input: positive 12-volt trigger pulse.
3. Operating frequency: greater than 200 kilohertz.
4. Power requirements: 12 volts D.C. at approximately 10 milliamperes.

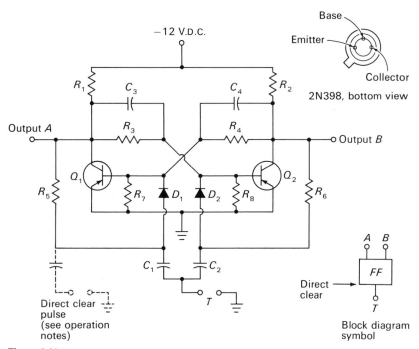

Figure 8.10
Circuit diagram of bistable (flip-flop) multivibrator.

The circuit diagram is given in Figure 8.10. The parts list is as follows.

R_1, R_2. 1,000-ohm, ½-watt.
R_3, R_4, R_5, R_6. 10,000-ohm, ½-watt.
R_7, R_8. 47,000-ohm, ½-watt.

C_1, C_2. 270-picofarad, ceramic.
C_3, C_4. 100-picofarad, ceramic.
D_1, D_2. 1N914.
Q_1, Q_2. 2N1303.
Miscellaneous: Circuit foundation, input/output jacks, hardware,
   wire.

Functional Description

The flip-flop or bistable multivibrator circuit is characterized by its ability to remain in the same state indefinitely after a trigger input signal is applied to it. It is in effect a memory element, capable of maintaining either of two possible states.

The operation of the flip-flop may be understood by assuming that transistor Q_1 is conducting and transistor Q_2 is cut off. The base of Q_1 receives enough negative bias voltage from the collector of Q_2 to maintain conduction. Transistor Q_2 remains cut off because not enough base bias voltage is provided through the Q_1 collector to the Q_2 base circuit (with Q_1 conducting, sufficient collector voltage is not available). The potential at the cathode of diode D_2 will be more positive than that at the cathode of diode D_1. In this condition, diode D_2 will be more forward-biased than diode D_1.

If a positive pulse is applied at point T, through capacitors C_1 and C_2, it will be able to pass through D_2 more easily than through D_1. A positive pulse at the base of Q_2 will forward-bias Q_2, lowering the Q_2 collector voltage. With Q_2 "on," the base of Q_1 will be reverse-biased and the flip-flop will have changed its state. It will remain in this condition until another trigger pulse is applied at point T. The output appears across collector resistors R_1 and R_2. Output A is 180 degrees out of phase with output B.

Operation Notes

Flip-flop circuits for use in high-speed counting and switching circuits cannot be designed without consideration of details that are beyond the scope of this book Although the circuit described here can operate at high switching speeds, it will function most reliably at relay switching speeds.

A relay (Sigma 4F1000, or equivalent) may be substituted for either collector resistor, R_1 or R_2. If a relay is used, transient voltage protection should be included. A diode (type 1N4005, or equivalent) wired across the relay coil will protect the flip-flop semiconductors from damaging transient voltages.

The use of flip-flop circuits in the construction of a binary counter is

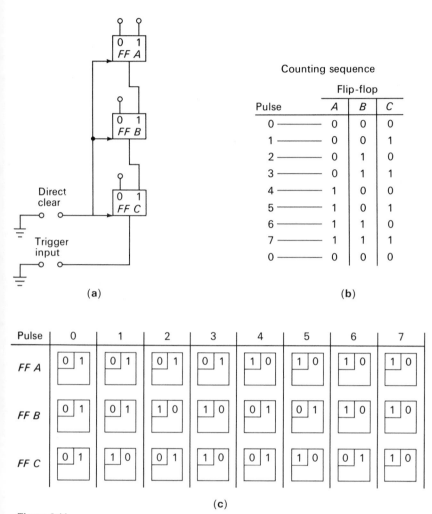

Counting sequence

Pulse	Flip-flop		
	A	B	C
0	0	0	0
1	0	0	1
2	0	1	0
3	0	1	1
4	1	0	0
5	1	0	1
6	1	1	0
7	1	1	1
0	0	0	0

(a)

(b)

(c)

Figure 8.11
Block diagram of binary counter (a); counting sequence (b);
and state of each flip-flop after each pulse (c).

shown in the block diagram, Figure 8.11, a. In this example, three flip-flops
are used to count sequentially from 0 to 7. The counting sequence (Figure
8.11, b) is listed next to the block diagram. The operation of the counter
may be understood by considering the state of the individual flip-flops
after each trigger pulse (Figure 8.11, c). The output, or count, is coded in
binary notation and is taken from the output terminal located in the upper
left-hand corner of each flip-flop. The sequence of operations can begin
only after all the flip-flops have been set to the same initial state by the

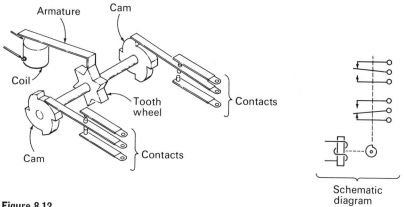

Figure 8.12
Basic sequencing relay.

application of a "direct clear" pulse. The direct clear pulse is a trigger pulse applied through a capacitor to the base of transistor Q_1 in the circuit diagram of Figure 8.10.

With the application of trigger input pulse "1," flip-flop C will change state. The process of changing state is known as "complementing." The other flip-flops will remain in their initial state, since no trigger pulse is present at their inputs. With the application of pulse "2," flip-flop C will again be complemented, a trigger pulse will appear at the input of flip-flop

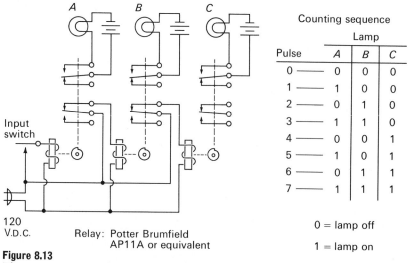

	Counting sequence		
		Lamp	
Pulse	A	B	C
0 ———	0	0	0
1 ———	1	0	0
2 ———	0	1	0
3 ———	1	1	0
4 ———	0	0	1
5 ———	1	0	1
6 ———	0	1	1
7 ———	1	1	1

0 = lamp off

1 = lamp on

120 V.D.C.   Relay: Potter Brumfield
AP11A or equivalent

Figure 8.13
Binary counter using sequencing relays.

B, and flip-flop B will be complemented. With the application of pulse "3," flip-flop C is complemented, and flip-flops B and A remain in the state shown in Figure 8.11, c. With the application of additional input trigger pulses, the count progresses as shown. After pulse "7" the cycle is repeated. Additional stages may be added as required to extend the counting range.

Certain types of relays known as sequencing relays may be used as flip-flop circuits in many slow-speed switching applications. The basic sequencing relay consists of a coil, armature, toothed wheel, cam, and contacts (Figure 8.12). Current flowing through the coil pulls the armature down, and the armature steps the toothed wheel one step. The cams, fixed to the same shaft as the toothed wheel, open and close the contacts. On application of a pulse, the contacts will change state and remain in the changed state position until a second pulse is delivered. An advantage of this type of flip-flop over the transistorized flip-flop is the permanent memory of the contact settings. With the transistorized flip-flop, current must be constantly applied to the circuit for it to remain in the same state. Figure 8.13 shows how sequencing relays may be interwired to form a binary counter. The method of advancing the count is similar to that employed in the transistorized counter.*

---

* A detailed analysis of flip-flop circuits, including design requirements for the circuit designer, can be found in Texas Instruments, Incorporated, *Transistor Circuit Design* (New York: McGraw-Hill, 1963), pp. 373–76.

LAMP AND RELAY DRIVERS

Applications and Specifications

1. Output indicators and activators to be used with digital-type programming equipment: type A, low-voltage transistorized lamp driver; type B, low-voltage transistorized lamp driver with latching function; type C, high-voltage neon lamp driver; type D, basic SCR latching circuit; type E, SCR lamp driver.

2. Operating requirements: types A and B, plus 12 volts D.C.; type C, minus 6 volts D.C.; types D and E, plus 3 to plus 6 volts D.C.

3. Power requirements: types A and B, 12 volts D.C.; type C, 6 volts D.C. and 100 volts D.C.; types D and E, 50 volts D.C. maximum at 1.5 amperes.

Functional Description

Type A (see Figure 8.14). The transistorized lamp driver operates exactly like the basic transistorized *AND* gate described earlier.

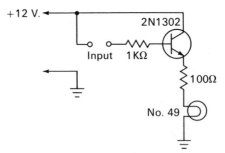

Figure 8.14
Low-voltage transistorized lamp driver.

Type B (see Figure 8.15). This circuit consists of the basic lamp driver, type A, plus a light-dependent resistor (LDR) latching circuit. Resistor R_2 and lamp I_2 are wired in parallel with indicator lamp I_1 and resistor R_1. Lamp I_2 is mounted closely to the LDR. When an input signal causes the transistor to conduct and lamps I_1 and I_2 to light, the resistance of the LDR is lowered and sufficient current flows through the LDR to maintain forward base bias on the transistor. The lamps will remain lit without an

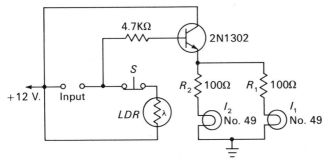

Figure 8.15
Low-voltage transistorized lamp driver with latching function.

input signal, or until the LDR base circuit is broken by pressing the momentary contact push-button switch S.

Type C (see Figure 8.16). With no input signal voltage, the transistor is cut off by the base bias voltage of R_2 and the positive 6-volt source. With the transistor cut off, lamp L_1, connected to the collector circuit, cannot receive enough voltage to light (lamp L_1 needs about 70 volts to light). A minus-6-volt input signal overcomes the transistor base bias and causes the transistor to conduct heavily, lighting the lamp.

Type D (see Figure 8.17). With no input signal applied to the SCR gate, the SCR cannot conduct enough to activate the load. A momentary positive gate pulse causes the SCR to saturate, and operating power is applied to the load. The SCR will conduct until operating power is removed. Resistor R assures reliable gating action and limits excessive currents.

Type E (see Figure 8.18). The basic SCR latching circuit may be modified into a high-power lamp driver by using a source of rectified A.C. power for the load and a low-voltage signal for the gate. The full-wave rectified current is not filtered, since the zero-voltage points of the waveforms are

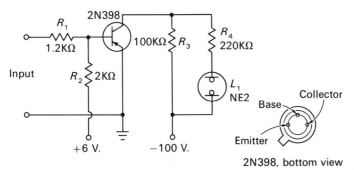

Figure 8.16
High-voltage neon lamp driver.

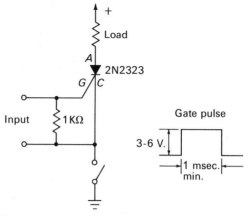

Figure 8.17
Basic SCR latching circuit.

used to extinguish the lamp load. With a signal applied to the gate lead, the SCR conducts, providing operating power to the load. When the signal is removed, the SCR will continue to conduct until the waveform, as measured across the SCR, approaches zero voltage and turns off the SCR, which cannot again conduct without the application of a gate signal.

Construction Notes

If D.C. driver circuits are to be used to operate relays, solenoids, or other inductive loads, a protective diode should be wired across the operating coil. The diode (type 1N4005, or equivalent) prevents excessive transient voltages from damaging the semiconductors in the circuit.

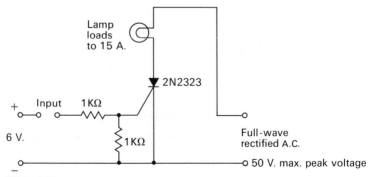

Figure 8.18
SCR lamp driver.

Different SCRs may be substituted for those listed. SCRs are commonly available with forward current ratings up to 35 amperes and forward blocking voltages up to 400 volts. Higher-powered SCRs must have adequate heat sinks.

INTEGRATED-CIRCUIT LOGIC GATES

Applications and Specifications

1. Type A, *AND/OR* gate; type B, *NAND/NOR* gate; type C, astable multivibrator; type D, monostable multivibrator.

2. Power requirements (all circuits): 1.5 to 4.5 volts D.C., battery or well-regulated power supply.

3. Input/output characteristics: specified by the integrated-circuit manufacturer as "units of drive" or "gain units." Typically, logic gates need three drive units for input, but deliver 16 drive units as output, and so can directly drive five logic-gate inputs. Table 8.1 lists the input and output drives for the four circuits.

Table 8.1
Input and Output of Integrated Circuits.

Circuit	Input (in drive units)	Output (in drive units)
AND/OR gate	3	16
NAND/NOR gate	3	16
Astable multivibrator	—	13
Monostable multivibrator	3	13

Functional Descriptions

The μL914 is an inexpensive IC consisting of four transistors and six resistors mounted on a single silicon chip and potted in an epoxy case only slightly larger than a conventional transistor packaged in a TO-5 case (see Figure 8.19). Operation of the IC may be understood by viewing the entire circuit as a conventional transistorized circuit, and not giving special consideration to the miniaturization. In the internal circuit of Fig. 8.19, the emitter leads of the four transistors are connected to a common point (lead 4). Each transistor contains a 450-ohm resistor in series with its base. These resistors prevent damaging currents, should the base leads be shorted directly to the positive supply voltage. Each transistor may be effectively removed from the circuit by shorting its base lead to its emitter lead.

Type A is an IC wired to function as an *AND/OR* logic gate (see Figure 8.20). The operation of IC logic-gating circuits requires the establishment

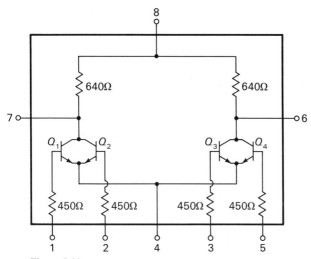

Figure 8.19
Internal circuit of μL914 IC.

of criteria for the input and output states. If, for example, a positive voltage signal is defined as a binary 1, while the absence of a positive voltage is defined as 0, the *OR* gate of Figure 8.20 will be formed.

With no input signals at input *A* and input *B*, transistors Q_1 and Q_2 will both be cut off and there will be a collector-to-ground voltage at lead 7. The voltage at lead 7 is applied through lead 3 to the base of Q_3. With transistor Q_3 on, the collector-to-ground voltage, or output voltage at

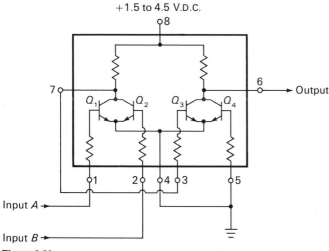

Figure 8.20
IC wired to function as an *AND/OR* logic gate.

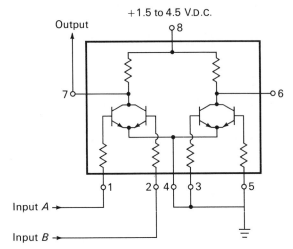

+1.5 to 4.5 V.D.C.

Figure 8.21
IC wired to function as a *NAND/NOR* logic gate.

lead 6, will be zero. With a positive signal applied to either input A or input B (or both together), the collector-to-ground voltage at lead 7 will be zero. The base of transistor Q_3 will have no voltage and the collector-to-ground voltage of transistor Q_3, as measured at lead 6, will be maximum, or binary 1. These conditions fulfill the requirements of a logical *OR* gate.

If the initial criteria are changed and a positive voltage signal is defined as a binary 0 while the absence of a positive voltage is defined as a binary 1, then the circuit of Figure 8.20 functions as an *AND* gate. With no input voltage at lead A nor at lead B (binary 1), there will be no output signal at lead 6 (binary 1). With an input signal voltage at lead A or at lead B or at both (binary 0) there will be an output signal at lead 6 (binary 0).

Type B is an IC wired to function as a *NAND/NOR* logic gate (see Figure 8.21). The same signal criteria may be applied to the *NOR* gate as was applied to the *OR* gate, and the same criteria to the *NAND* gate as to the *AND* gate.

With a positive signal voltage defined as a binary 1 and with input signals applied to either input A or input B or both, the collector voltage at lead 7 will be zero (binary 0). With no inputs, the lead 7 voltage will be maximum (binary 1). These conditions fulfill the requirements of the logical *NOR* gate.

With a positive signal voltage defined as binary 0 and with no signal voltage defined as binary 1, and with input signals at either input A or input B, there will be no signal voltage at lead 7. With no signal voltages, there will be a signal voltage at lead 7. These conditions fulfill the requirements of the logical *NAND* gate.

$$C = C_1 = C_2$$

$C\ \mu\text{fd}$	Frequency range (hertz)
10	10–100
1	100–1,000
.1	1,000–10,000
.01	10,000–100,000

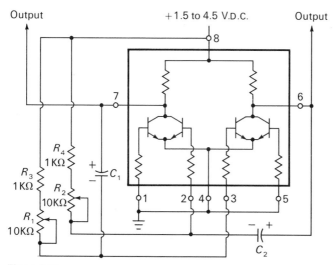

Figure 8.22
IC wired to function as an astable multivibrator.

Type C is an IC wired to function as an astable multivibrator (see Figure 8.22). The basic operation of the astable multivibrator is discussed on p. 148. The operation of the circuit of Figure 8.22 is identical to the basic astable multivibrator. Capacitors C_1 and C_2, together with controls R_1 and R_2, determine the operating frequency of the multivibrator. By selecting various values of capacitance for C_1 and C_2, various operating frequencies are possible; controls R_1 and R_2 determine the precise frequency within the desired range.

Type D is an IC wired to function as a monostable multivibrator (see Figure 8.23). The basic operation of the monostable multivibrator is discussed in Chapter 6. The circuit of Figure 8.23 is identical to the basic monostable multivibrator circuit, except that the IC circuit lacks the triggering circuit. The identical triggering circuit may be used with the IC, if the circuit is to be used with irregular inputs, such as mechanical switch closures.

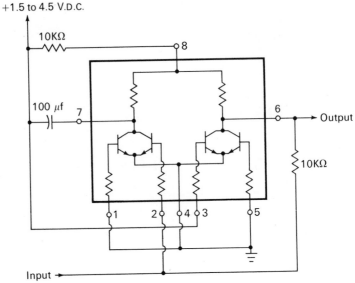

+1.5 to 4.5 V.D.C.

10KΩ

100 μf

Output

10KΩ

Input →

Figure 8.23
IC wired to function as a monostable multivibrator.

The output pulse width is determined by the value of capacitor C (approximately 200 microfarads per second of pulse duration). With the value shown, the output pulse is approximately one-half second in duration.

Construction Notes

The IC may either be mounted in a socket or may be lead-mounted by carefully bending each lead, one at a time, at right angles to the case, and leaving approximately $\frac{1}{16}$-inch clearance between the case and the bend. The leads may then be separately soldered to stand-off terminals mounted on the chassis.

The Fairchild μL914 integrated circuit is manufactured with two case designs. The lead designations are illustrated in Figure 8.24. Both cases contain exactly the same internal circuitry.

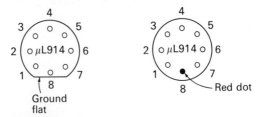

Ground flat

Red dot

Figure 8.24
Bottom view of μL914 IC, showing two methods of lead designation.

9

Time-Delay Circuits

All timing circuits require an accurate and consistent time base in order to be reliable. In ordinary electric clocks the time base is maintained by the electric-power companies at a precise 60-hertz frequency. The time base commonly used in electronic timing circuits is developed by a resistor, a capacitor, and some type of electronic sensing element. In a typical R-C time-delay circuit, a capacitor is charged exponentially through a resistor. At some point during the charging period, there is enough voltage across the capacitor to activate the sensing element, causing it to conduct sufficiently to activate a relay or other control element.

Time-delay circuits form the basic element in a variety of timing devices, including clocks used in programming circuits, elapsed-time indicators, oscillators, counters, and pulse formers. Often the output of the basic time-delay circuit initiates repeated timing cycles, forming a circuit known as a relaxation oscillator. Other time-delay circuits use thermistors, and depend on circuit heating for a time delay before sensing elements are activated; these types of circuits must be returned to the initial operating temperature before the same time delay can be repeated.

BASIC TIME-DELAY RELAY

The basic R-C time-delay relay is shown in Figure 9.1. In switch position A, the capacitor can charge through a current-limiting resistor, which

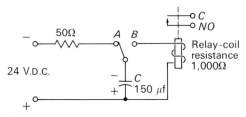

Figure 9.1
Basic R-C time-delay relay.

prevents surge currents from damaging the capacitor. (The charging of a capacitor was discussed briefly in Chapter 7, in reference to oscillator circuits; that discussion is equally applicable to time-delay circuits.) When the switch is thrown to position B, the fully charged capacitor can discharge through the relay coil, which in this circuit functions both as the sensing element and as the control element.

The capacitor discharge time can be approximated as

time constant (in seconds) = R (in ohms) $\times$ C (in farads).

The "time constant" is the time it takes the voltage across the capacitor to drop to approximately one-third of its starting value. With the circuit values of Figure 9.1, the time constant is

$$150 \times 10^{-6} \text{ farads} \times 1{,}000 \text{ ohms} = 150 \times 10^{-3} = 0.15 \text{ seconds.}$$

At the end of 0.15 seconds, the voltage across the capacitor has dropped about 16 volts, leaving about 8 volts across the relay coil. A second period of 0.15 seconds will result in a voltage drop of approximately ⅔ of 8 volts, or about 5.3 volts, leaving 2.7 volts across the relay coil. If the relay opens when the voltage across its coil is 4 volts, then the time delay before release will be between 0.15 and 0.3 seconds. In order to lengthen the time delay, a larger capacity capacitor, a relay with a larger coil resistance, or an additional resistor in series with the relay coil, may be employed.

The accuracy of a relay-capacitor timer depends on the operating characteristics of the relay, the regulation of the source voltage, and the tolerance and reliability of the timing capacitors. Of these factors, the relay operating characteristics are the most difficult to control. Because of mechanical wear and contact chattering, the relay activation and release times may vary several milliseconds over the course of several minutes of

repeated use. Although these variations may be acceptable in many operations, they are unacceptable in timing operations that involve simultaneous
presentation of different timing signals or the creation of a time base from
which functions are being repeated, such as a clock circuit used to drive
digital switching circuits.

For increased accuracy (2 per cent repeatability or better), an electronic
sensing element must be used in place of the relay. The factors affecting
the accuracy of an electronic timing circuit include regulation of the source
voltage, maintenance of a stable operating temperature, the stability of the
sensing circuit, and capacitor tolerance and reliability. Each of these
factors may be controlled to some degree. Well-regulated power supplies
may be employed as power sources, operating temperatures may be held
relatively constant by using thermistors and diodes in the sensing circuit,
and high-grade, close-tolerance capacitors may be employed.

The basic time-delay relay shown in Figure 9.1 produces its timing
interval as a control signal (relay closure) as soon as the switch is closed.
The relay remains activated for the timed interval and then drops out.
There are other time-delay relays that, when its switch is closed, remain
inactive until a predetermined time interval has elapsed, and then activate
a relay or other control element, which remains activated until the control
circuit is broken. Both techniques are used in control applications,
although the latter type is more common, and will be used in the remaining
circuits to be described in this chapter. The basic principle, however,
remains the same as in the time-delay relay shown in Figure 9.1; the
charging of a capacitor through a resistor is used to create a time interval
for a sensing and control stage.

TIMING CAPACITORS

The timing capacitor should be selected carefully. Accurate *R-C* timing
circuits need capacitors that operate with a high degree of stability, that is,
reliably and with a minimum of power losses. A major source of power
loss is small currents leaking through the dielectric material separating one
plate of the capacitor from the other. Excessive leakage may cause heating,
which could destroy the insulating properties of the dielectric, reducing or
eliminating the capacitance of the capacitor.

Electrolytic capacitors are commonly used in timing circuits because
they have greater capacity for a given size than most other capacitor types.
Electrolytic capacitors, when used for the first time or after having been
out of service for some time, should be slowly "formed." Forming is
necessary to insure the development of a stable oxide layer, which functions
as the dielectric. To "form" an electrolytic capacitor, a small voltage is
applied to the capacitor (observing the polarity marked on the capacitor

case). The voltage is gradually raised until the operating voltage is reached. The entire process may take several hours, depending on the capacitance and the working voltage of the capacitor.

In addition to electrolytic capacitors, other, more costly types of capacitors are used in *R-C* timing circuits. High-stability, low-loss tantalum electrolytic capacitors are available with capacitance ratings up to 40 microfarads at 50 working volts. (Capacities up to approximately 400 microfarads are available at lower working voltages.) Where lower capacity ratings are required, plastic-film capacitors may be employed. These are very high-stability, low-loss capacitors with ratings up to about 1 micro-farad with voltage ratings up to 200 volts. When using capacitors, one should not exceed the maximum working voltage specified by the manu-facturer and usually marked on the case (working volts D.C. or W.V.D.C.) but lower voltages may be used. In oscillator circuits, the peak voltage (not the average or RMS voltage) must not be exceeded. Also, the polarity of electrolytic capacitors must be carefully observed, or one will destroy the carefully formed oxide layer and the capacitor.

BASIC UNIJUNCTION TRANSISTOR TIME-DELAY RELAY

The circuit diagram is given in Figure 9.2. When power is applied through SW, capacitor C takes on a charge through R_4 and the variable resistor R_1. Initially the relay is inactive, because the current through the unijunction transistor is cut off; that is, there is insufficient current flow to activate the relay. The unijunction transistor remains cut off until its "emitter peak point voltage" (that is, the voltage between the emitter and ground) reaches approximately 14 volts. At that point the emitter becomes forward-biased, the unijunction transistor is permitted to conduct, and the relay is activated. In this circuit the relay contacts are used to latch the relay. The resistors R_2 and R_3 are current-limiting resistors, while the diode D is used to protect the transistor from harmful voltage transients.

The time delay of this circuit is determined by the setting of R_1. There is approximately one second of delay for each 10,000 ohms of resistance. For accurate calibration, a timing standard should be compared with various settings of R_1. Additional accuracy is possible by using a precision-calibrated potentiometer (helipot) for R_1.

The relay specified in Figure 9.2 is a special, fast-acting, low-voltage type that is needed to insure reliable contact closure. An ordinary relay may instead be used if an SCR is used as the control element.

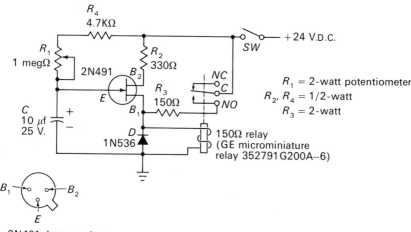

Figure 9.2
Basic unijunction transistor time-delay relay
(courtesy of General Electric Company).

UNIJUNCTION TRANSISTOR SOLID STATE TIME DELAY

The circuit diagram is given in Figure 9.3. This circuit illustrates how timing control can be achieved without using a relay. The timing or passive portion of the circuit operates like the circuit of Figure 9.2. The zener diode, D_1, acts as a voltage reference for the timing circuit and limits the voltage applied across the circuit to 18 volts. At the end of the timing period, determined by the setting of R_1, the unijunction transistor triggers the SCR. The SCR functions as a very sensitive latching relay. When its gate lead is triggered, the SCR conducts heavily and permits the entire line voltage, minus approximately 1 volt, to be delivered to the external load. The load current is limited by the rating of the SCR. With the C20F shown in Figure 9.3, the load-current limit is 6 amperes. In this circuit, time delays from 0.4 milliseconds to 1 minute are possible, using various values of capacitance as well as various settings for the timing resistor, R_1. As in the previous circuit, there is approximately one second of delay for each 10 kilohms of resistance. The values of capacitance may vary from about 0.2 microfarads, for timing intervals from 0.4 milliseconds to 0.1 seconds, to 100 microfarads, for timing intervals from 0.2 seconds to approximately 1 minute. For greatest accuracy (1 per cent or better repeatability), a precision-calibrated resistor (helipot) should be used for R_1 and a low-leakage capacitor used for C_1. As with the circuit of Figure 9.2, the current must be removed from the circuit to reset the circuit.

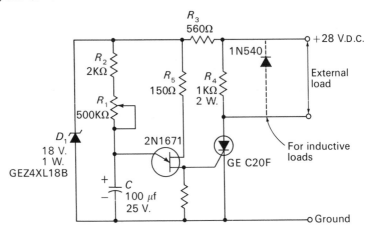

Figure 9.3
Circuit diagram of unijunction transistor solid state
time-delay relay (courtesy of General Electric Company).

SILICON-CONTROLLED RECTIFIER, 117-VOLT A.C. TIME-DELAY RELAY

The circuit diagram is given in Figure 9.4. In this time-delay relay, a controlled rectifier, SCR_1, functions both as the sensing element and as a sensitive electronic relay that supplies current to energize an output relay or other load. With push-button switch S_1 in the reset position, timing capacitor C_1 charges to the negative peak supply voltage through R_1 and D_1. When S_1 is released, that is, set to the "time" position, diode D_2 begins to conduct and capacitor C_1 begins to discharge toward the positive peak supply voltage through R_3, R_2, and D_2. The rate of discharge is determined by the setting of potentiometer R_2. When C_1 reaches about plus 2 volts, the control rectifier is triggered on. Capacitor C_2 and diode D_3 function as a pulse expander, permitting the SCR to deliver a pulse long enough to enable a load relay to latch on. Capacitor C_2 rapidly charges to a positive potential before C_1 discharges into the SCR gate. When C_1 has discharged, C_2 continues to discharge into the SCR gate. Diode D_3 is reverse-biased and prevents C_2 from discharging toward capacitor C_1.

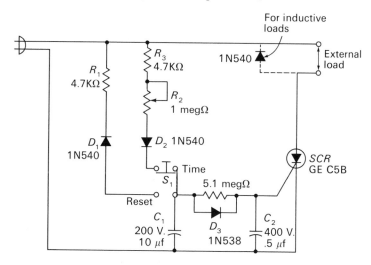

C_1 = Nonpolarized metalized capacitor (Cornell-Dubilier AM2–106, or equivalent)

Figure 9.4
Silicon-controlled rectifier time-delay relay (courtesy of General Electric Company).

THERMISTOR TIME-DELAY RELAY

The circuit diagram is given in Figure 9.5. This circuit can provide a time delay of from 0.5 to 15 seconds, depending on the setting of potentiometer R_1. The time delay is created by the circuit heating and the subsequent change in the resistance of the thermistor. The more current that is permitted to flow through R_1 and the thermistor, the faster the thermistor heats. As the thermistor heats, its internal resistance drops until at some point sufficient current flows to activate the control relay. Once the relay is activated, it remains on until switch SW_1 is opened.

There are many commercial variations of this type of thermal time delay. They are sometimes used to switch on the plate circuits in vacuum tubes after allowing time for the tube filaments to heat up. In certain types of Thyraton tubes, for example, the tubes may be damaged unless sufficient warm-up time is allowed before the tube is permitted to conduct.

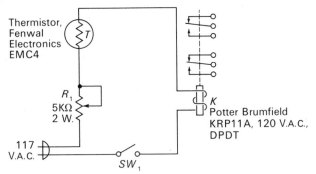

Figure 9.5
Circuit diagram for thermistor time-delay relay.

FURTHER INFORMATION

In addition to the time-delay circuits presented in this chapter, accurate and reliable electronic timing circuits may be constructed using multi-vibrator circuits (see pp. 115 and 148), various relaxation-oscillator circuits (see p. 144) and Schmitt trigger circuits (see p. 113). Each of these circuits operates on the same R-C timing principles described in the introduction to this chapter. Accordingly, by using various valves of R and C, one can adjust the circuit timing interval to fit a specific application.

An example of an unusual relaxation-oscillator circuit using a photocell and a neon bulb is shown in Figure 9.6. The circuit is designed to operate from direct current obtained directly from the 117-volt A.C. power line through a bridge rectifier and filter capacitor. This direct current is used to charge capacitor C through timing control R_1. When the charge on C is sufficient to ignite the neon bulb, NE-2, the light-dependent resistor, LDR, which is mounted closely to the neon bulb, reduces the resistance sufficiently to activate relay K. When the NE-2 discharges C, the resistance of the LDR rises, the relay opens, and the timing cycle again begins. When constructing this circuit, care must be exercised in mounting the LDR closely to the NE-2 and in preventing extraneous light from striking the LDR. In this circuit, the time interval between pulses is adjustable between approximately 30 seconds and 3 minutes.

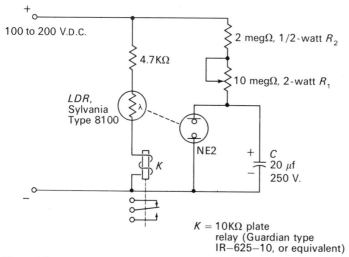

Figure 9.6
Circuit diagram for relaxation oscillator using a photocell and neon bulb.

10

Audio Circuits

Circuits that amplify audio signals accurately and reliably by means of transistors have been developed and refined within the past few years, and circuit designers have succeeded in packaging high-gain, high-quality circuits in small, compact modules. The modular design concept has also led to the availability of preassembled, low-cost, audio preamplifier and amplifier modules containing no power supply or extraneous hardware (see Figure 10.1). Many require only a power source, input signal, and output load to function. They are presently available with power ratings of up to 20 watts and frequency responses between 20 and 20,000 hertz. Table 10.1 gives the operating specifications of several popular audio amplifier modules.

In addition to modular audio amplifiers, low-cost integrated-circuit audio amplifiers are available; these are complete multistage direct-coupled amplifiers, packaged in standard transistor-size cases, that amplify audio frequencies with a minimum of distortion. These ICs require only simple input and output impedance matching networks, together with bias-stabilizing networks. At present these ICs are available with power

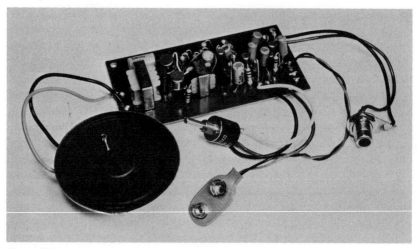

Figure 10.1
Popular, low-cost audio-amplifier module.

ratings of up to about 2 watts, with more power easily added by coupling transistor output stages to the IC.

Examples of IC audio amplifiers are included in this chapter to illustrate methods of impedance matching and interwiring. The low cost, availability, and efficiency of modular audio amplifiers and preamplifiers preclude the presentation of amplifier circuits in this book.

AMPLIFIER CHARACTERISTICS

The characteristics by which amplifiers and other audio circuits are measured have been standardized for operating power, frequency response, and distortion. Audio power is commonly specified in watts of r.m.s. sine-wave power.* Sine-wave power is calculated as

$$P = \frac{E^2}{Z},$$

where P is power in watts, E is applied sine-wave voltage (r.m.s.), and Z is the impedance across which the power is being developed.

The decibel (dB) is convenient for specifying the increase in gain (ratio

---

* Manufacturers of audio (high-fidelity) equipment may rate their equipment by any of the following standards: sine-wave power, peak sine-wave power, I.H.F. (Institute of High Fidelity) music power, or E.I.A. (Electronics Industries Association) music power. Sine-wave power is discussed in the text; peak sine-wave power is exactly twice sine-wave power. I.H.F. music power is sine-wave power at 1 per cent THD. E.I.A. music power is sine-wave power at 5 per cent THD.

Table 10.1
Specifications of Popular Transistorized Amplifier Modules.

Manufacturer	Module	Model no.	D.C. Power supply voltage	Power (watts, 8-ohm load)	Input impedance (ohms)	Output impedance (ohms)	Frequency response (hertz)
Amperex	Amplifier	PCA-1-9	8	1	600 K	8-16	100-50 K
	Preamp. and amplifier	PCA-3B-18-1	18	4	1,000 K	8	45-20 K
	Amplifier	PCA-8-36	36	20	1,000 K	8	20-40 K
	Stereo preamp.	PCA-7B-18	14-18	*	20 K	10 K	*
G-C Audio-Tex	Amplifier	32-800	6	2	*	8	*
International Rectifier (ECT-O-PAX)	Amplifier	EA-305	6	260 mW	*	8	250-15 K
Lafayette	Amplifier	99C9037	9	360 mW	2 K	8	*
	Amplifier	99C9132	9	3	2 K	8	150-10 K
	Solid state amplifier	19C0107	6	*	*	8	*

* Not specified by manufacturer.

of output to input) of a stage or stages in an amplifier. The decibel is the logarithm of the ratio between two voltages, currents, or powers. The formulas used to calculate decibels in audio calculations are:

$$dB = 10 \log \frac{P_2}{P_1} \quad \text{(for power calculations)},$$

$$dB = 20 \log \frac{E_2}{E_1} \quad \text{(for voltage calculations)},$$

where P_2 is the power output in watts, P_1 is the power input in watts, E_2 is the signal voltage output, and E_1 is the signal voltage input.

Use of the decibel depends on having a reference level either stated or implied. In audio applications, a common reference level for measuring signal voltages is 0.775 volts. When this voltage is applied to a resistance of 600 ohms, one milliwatt of energy is dissipated. Most volt-ohm meters or vacuum-tube voltmeters contain a dB scale referenced to a 1-milliwatt, 600-ohm level. When a meter is used to measure decibels in circuits where the impedance is other than 600 ohms, corrections must be made in order to reference the measured reading to the 600-ohm, 1-milliwatt meter scale. The correction factor to be added to the meter reading is $10 \log_{10} (600/E)$, E being the impedance across which measurements are being made.

When the low frequency response of an audio circuit is specified, for example, as minus 3 dB at 20 hertz, as measured with a standard dB meter, this specification means that the power output at 20 hertz has dropped by 50 per cent from that power output measured throughout the audio range; for every 3 dB decrease, there is a halving of the power level.

One of the most significant measurements of distortion in an audio circuit is that of harmonic distortion, which represents the inability of an amplifier to reproduce a signal perfectly. When a single sinusoidal frequency is applied to an amplifier that, because of its inherent characteristics, cannot reproduce this frequency perfectly, the output will contain overtones of the fundamental frequency that are not present at the input. The magnitude of these overtones or harmonics, expressed as a percentage of the original frequency, may be measured with a special wave analyzer. The total harmonic distortion (THD) is the sum of the square roots of all the harmonic components produced in the amplification of a sine wave. Modern transistor amplifiers produce a total harmonic distortion of 1 per cent or less.

A second important measure of amplifier distortion is known as inter-modulation distortion (IM). Intermodulation distortion is the production of spurious frequencies within the amplifier by the interaction of two signals. IM distortion is measured by feeding two different audio frequencies into the amplifier, and filtering one frequency out of the output. If the amplitude of the second frequency fluctuates, IM distortion is present. The second frequency is then also filtered out, and any remaining audio voltage is measured as a percentage of the output waveform before filtering. The frequencies used in making IM tests should be given in the IM specifications of an amplifier. Modern transistor amplifiers produce an IM distortion of 1.5 per cent or less.

BASIC THREE-TRANSISTOR AUDIO AMPLIFIER

Applications and Specifications

1. Extremely simple audio amplifier, useful for voice amplification where quality of sound reproduction is not critical.
2. Frequency response: approximately 100 to 5,000 cycles.
3. Output power with approximately 1-volt input: 140 milliwatts with a 4-ohm speaker, 3 milliwatts with a 16-ohm speaker.
4. Operating power: 1.5 to 6 volts.

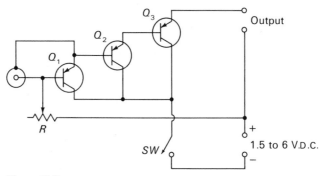

Figure 10.2
Circuit diagram for basic three-transistor audio amplifier.

The circuit diagram is given in Figure 10.2. The parts list is as follows.

R. 250,000-ohm, 2-watt potentiometer.
Q_1, Q_2. 2N109.
Q_3. 2N2869.
SW. SPST.
Miscellaneous: Foundation board, terminals, wire, hardware.

Functional Description

Three transistors are direct-coupled. The biasing of the circuit is provided by the gain control, R, which sets the gain for the entire circuit.

Construction Notes

A variety of audio transistors will work well with this circuit; Q_1 and Q_2 should be small-signal transistors, and Q_3 a power transistor which does not need a heat sink.

Operation Notes

1. Adjust gain from the high-resistance side of R. The response is most efficient when R is set to the high-resistance side of the point where distortion occurs.

2. An electrolytic capacitor (approximately 5 microfarads) should be placed in series with the input when driving the amplifier from a microphone (dynamic or magnetic).

3. For output loads below 8 ohms, use 4.5 volts or less for operating voltage.

4. This amplifier is useful as a visual indicator or as a relay control amplifier for low-level inputs. A standard pilot lamp or a sensitive miniature relay may be used for the output instead of the speaker.

AUDIO MIXER

Applications and Specifications

1. Mixes two or more audio inputs into a single output.
2. Separate gain control for each input.
3. Input impedance: adjustable, approximately 2 megohms maximum.
4. Circuit gain: approximately 5 dB. Output should be coupled to low-level preamplifier input.
5. Frequency response: 20 to 20,000 hertz, ±1 dB.
6. Total harmonic distortion: less than 1 per cent.
7. Power requirements: 12 to 24 volts D.C., battery or well-regulated power supply.

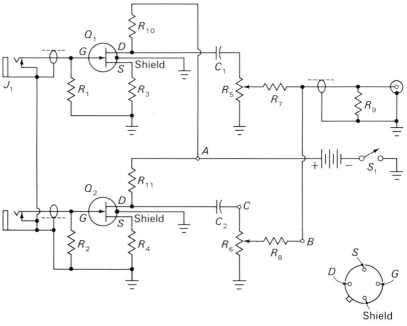

Figure 10.3
Circuit diagram of audio mixer.

The circuit diagram is given in Figure 10.3. The parts list is as follows.

R_1, R_2. 2.2-megohm (see Construction Notes).

R_3, R_4. 6,800-ohm.

R_5, R_6. 500,000-ohm, audio-taper potentiometer.

R_7, R_8. 100,000-ohm, ½-watt, 10%.

R_9. 100,000-ohm, ½-watt, 10%.

R_{10}, R_{11}. 560-ohm.

C_1, C_2. 0.1-microfarad, ceramic disc.

Q_1, Q_2. HEP-801.

J_1, J_2. Shorting-type phone jacks.

J_3. Phono jack.

Miscellaneous: Shielded cable, circuit foundation, switch, battery, wire, hardware.

Functional Description

Two field-effect transistors (FET), Q_1 and Q_2, are the mixing circuit element. The FET has an input impedance of several megohms. The exact value of the impedance is determined by gate resistors R_1 and R_2. The individual FET outputs are capacitor-coupled to individual level controls R_5 and R_6. The mixed signal is then coupled to the output jack. The remaining resistors are used to establish circuit operating levels, while capacitors C_1 and C_2 couple the signals to the level controls.

Construction Notes

For maximum performance, resistors R_1, R_2, R_3, and R_4 should be carbon film, low-noise types (Texas Instrument type CR, or equivalent).

When using shielded cable between J_1 and the FET gate lead, connect the shield at both ends. The shield for J_2 (and subsequent stages, if added) should be connected at the FET end, but not at the jack end, to avoid noise-producing ground loops. The unconnected shield should be taped to prevent short circuits. A ground buss should be used for connecting the FET shields, the volume control ground, and the shielded cable shield.

Additional input stages may be added to the original circuit. The positive power connection is to point A and the signal to be mixed is connected to point B.

Up to 5 dB of additional gain may be obtained (with some increase in distortion) by shunting resistors R_3 and R_4 with 30-microfarad electrolytic capacitors.

A closed-circuit phone jack may be inserted at point C to accept high-level inputs, such as those from a tuner or a preamplifier. The jack should be wired so that a phone plug inserted in the jack will disconnect the FET from the level control.

WHITE-NOISE GENERATOR

Applications and Specifications

1. Provides a source of white noise for masking environmental noises and for various acoustical experiments.
2. Covers the frequency spectrum between 20 and 20,000 hertz, ±2 dB.
3. Output: between 500 and 2,500 microvolts across a 100,000-ohm lead.
4. Current drain: between 1.0 and 1.5 milliamperes.
5. Power requirements: 6 to 24 volts D.C.

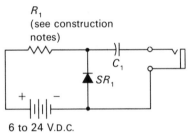

R_1
(see construction notes)

C_1

SR_1

+ −

6 to 24 V.D.C.

Figure 10.4
Circuit diagram of white-noise generator.

The circuit diagram is given in Figure 10.4. The parts list is as follows.

R_1. See construction notes.
C_1. 0.22-microfarad, ceramic.
SR_1. Solitron SD1-W white-noise diode.
Miscellaneous: Battery or well-regulated power supply, output jack, circuit foundation, wire, hardware.

Functional Description

The basic white-noise generator may be connected to the high-level (100,000 ohms or greater) input of a preamplifier to provide operating signals to a power amplifier. The white-noise diode, SR_1, is a specially constructed device that goes into avalanche or zener operation at a specific voltage. In the basic generator the output is coupled through a capacitor,

C_1, to a preamplifier. The value needed for resistor R_1 is determined by the particular diode obtained from the manufacturer (values between 22 and 100 kilohms are typically required).

Construction Notes

The value of R_1 is determined by calculating how much resistance will be needed for the specified current, given the classifying voltage of the diode. This is a simple Ohm's law calculation, $V = IR$, where V is the operating voltage of the circuit, I is the specified current, and R is the required value of resistance. The desired operating voltage should be specified when ordering the SD1-W diode.

Operation Note

The white-noise generator is a high-impedance device and should not be connected across a load much smaller than 100,000 ohms.*

---

* This circuit is abstracted from Lon Edwards, "White Noise, Its Nature, Generation, and Applications," *Electronics World* (Nov. 1962), pp. 40–42. An interesting variation of this circuit is described in Charles T. Tart, "An Inexpensive Masking Noise Generator: Monaural or Stereo," *Journal of Psychophysiology*, **2** (1966), 170–72. The SD1-W white-noise diode is available from Solitron Devices, Inc., 500 Livingston Street, Norwood, New Jersey.

METRONOME

Applications and Specifications

1. Source of adjustable metronomic timing beats produced in a self-contained loudspeaker.
2. Beat rate: adjustable from approximately 40 to 220 beats per minute.
3. Power requirements: 24 volts D.C. at 4 milliamperes.
4. May be coupled to an external audio amplifier for greater volume.
5. Accuracy is comparable with standard metronomes.

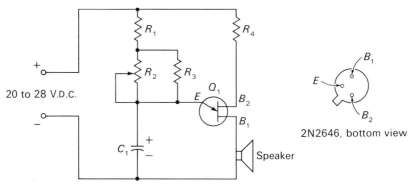

Figure 10.5
Circuit diagram of metronome.

The circuit diagram is given in Figure 10.5. The parts list is as follows.

R_1. 22,000-ohm.
R_2. 150,000-ohm, 2-watt potentiometer (log taper).
R_3. 430,000-ohm, ½-watt.
R_4. 330-ohm, ½-watt.
C_1. 10-microfarad, 50-volt, electrolytic.
Q_1. 2N2646.
Miscellaneous: 8-ohm small speaker, circuit foundation, battery or power supply, hardware, wire.

Functional Description

Capacitor C_1 is charged through resistor R_1 and potentiometer R_2 in parallel with R_3. When the charge on the capacitor reaches the "emitter peak-point voltage" of the unijunction transistor, Q_1, the emitter of Q_1 becomes forward-biased and the capacitor discharges through the B_1 lead of Q_1 and through the speaker, producing a distinct beat. When the capacitor is discharged, the emitter of Q_1 is reverse-biased (nonconducting), and the capacitor is prepared to receive another charge through the resistors. Potentiometer R_2 determines the charging time of the capacitor, while R_1 and R_3 determine the high-rate limit and the low-rate limit, respectively.

Construction Notes

Pulse rates can be changed by replacing C_1. Increasing R_2 will lengthen the period between beats; decreasing C_1 will shorten the period between beats. The circuit can be calibrated by using a stop watch for the slow-to-medium beats, and by comparing the beats with a calibrated metronome for the faster rates. An audio amplifier can also be used, rather than just the speaker, to provide louder beats.

VOICE-ACTIVATED RELAY

Applications and Specifications

1. High-sensitivity voice-activated relay. Normal conversation at a distance of about 25 feet from the microphone will activate the relay.

2. Activation time of relay is approximately 50 milliseconds following detection of noise.

3. Release time of relay following cessation of noise is adjustable from approximately 50 milliseconds to over 1 second (see Construction Notes).

4. Power requirements: 9 volts D.C. from battery or power supply.

5. Relay pull-in current: approximately 12 milliamperes.

6. Relay idle current: approximately 3 milliamperes.

The circuit diagram is given in Figure 10.6. The parts list is as follows.

R_1. 1-megohm.
R_2. 270,000-ohm.
R_3, R_7, R_8. 2,700-ohm.
R_4, R_9. 560-ohm.
R_{12}. 47,000-ohm.
R_5. 1,000-ohm.
R_6. 15,000-ohm.
R_{10}. 33,000-ohm.
R_{11}. 22,000-ohm.

R_{13}. 1-megohm potentiometer, with switch S_1.
C_1. 0.1-microfarad, ceramic.
C_2, C_4, C_6, C_8. 100-microfarad, 12-volt D.C. electrolytic.
C_3, C_5, C_7. 10-microfarad, 12-volt D.C., electrolytic.
D_1. 1N34.
Q_1, Q_2, Q_3, Q_5. 2N2925.
Q_4. 2N404.
K_1. Relay with pull-in current of 7 milliamperes D.C. (Sigma 11F-1000-G1516, or equivalent).
J_1, J_2. Shielded 4-pin miniature chassis connector (Amphenol 78-PCG4, or equivalent).
P_1, P_2. Shielded 4-pin miniature male cable connector (Amphenol 91-MPM42, or equivalent, to match J_1 and J_2).
Miscellaneous: Circuit foundation (printed circuit), battery or well-regulated power supply, hardware, wire.

Functional Description

A portion of the sound picked up by a crystal or ceramic microphone is passed through a sensitivity control to a high-gain audio amplifier stage, Q_1 and Q_2. The signal is then directed to the relay control stage. The relay

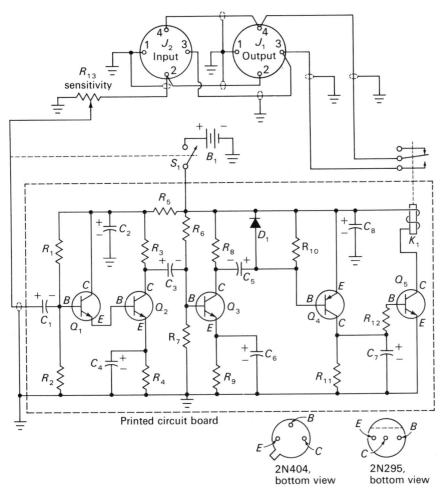

Figure 10.6
Circuit diagram for voice-activated relay.

contacts are wired into a connector, which is used both to activate the equipment being controlled and to pass the voice signal to the equipment.

Construction Notes

Layout of parts is important to prevent feedback loops and parasitic oscillations. If the printed circuit is not employed, the layout of parts should approximate the layout of the printed circuit (see Figure 10.7). Shielded cable should be used wherever indicated on the circuit diagram, with the shields connected only at one end. Hold-in time of the relay may

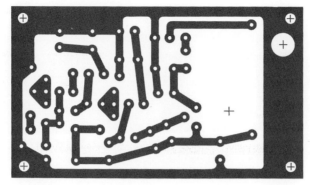

Printed circuit layout

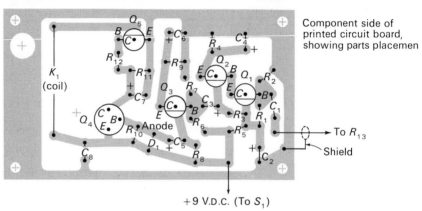

Component side of printed circuit board, showing parts placemen

+9 V.D.C. (To S_1)

Figure 10.7
Printed circuit board (actual size) and parts placement for voice-activated relay.

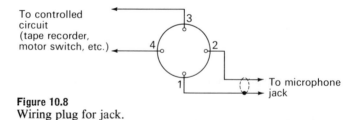

Figure 10.8
Wiring plug for jack.

be varied by selecting suitable values for capacitor C_7. With C_7 equal to 10 microfarads, the hold-in time is approximately one half-second. Increasing the value of C_7 increases the hold-in time; decreasing the value of C_7 decreases the hold-in time. A plug for jack J_1 should be wired as indicated in Figure 10.8.*

---

* This circuit is abstracted from Charles Caringella, "Build the Trans-Vox," *Popular Electronics*, **26** (Jan. 1967), 57–61. Information on how to obtain a complete kit of components can be found in that article.

INTEGRATED CIRCUIT AUDIO AMPLIFIER

Applications and Specifications

1. Two IC audio amplifiers, both using the inexpensive RCA CA3020 IC. The first circuit uses the IC alone to drive a loudspeaker, by means of an output transformer. The second circuit uses the IC to drive a higher-power audio amplifier.

2. Both IC amplifiers may be used as a general purpose audio amplifier, voice-operated relay, booster amplifier, signal tracer, preamplifier, intercom, or recorder monitor.

3. Input impedance (both circuits): 50 kilohms.

4. Power output, IC directly coupled to output: 545 milliwatts.

5. Power output, IC with output transistor stage: 4 watts.

6. Total harmonic distortion (both circuits): approximately 2 per cent.

7. Input voltage for rated power output (both circuits): 45 millivolts.

8. Power requirements (both circuits): 6 to 9 volts D.C. plus 12 volts D.C. for 4-watt amplifier.

9. Frequency response at rated power output (both circuits): approximately 40 to 20,000 hertz.

The circuit diagram is given in Figure 10.9. The parts list for the basic audio amplifier is as follows.

R_1. 510-ohm, ½-watt.
R_2. 5-kilohm, 2-watt potentiometer.
R_3. 0.62-ohm, ½-watt.
C_1. 1-microfarad, 15-volt, electrolytic.
C_2. 0.1-microfarad, paper.
C_3. 0.01-microfarad, ceramic.
C_4. 1-microfarad, 15-volt, electrolytic.
T_1. Transistor output transformer: primary, 125 ohms centertap; secondary, to match load impedance (Argonne type 176, or equivalent for 8-ohm speaker).
Miscellaneous: Input jack, circuit foundation, output socket, IC heat sink, wire, hardware.

The parts needed for the additional 4-watt output stage are as follows.

R_3, R_6. 1.1-ohm, ½-watt.
R_4. 220-ohm, ½-watt.

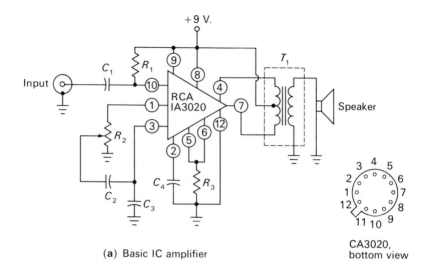

(a) Basic IC amplifier

CA3020,
bottom view

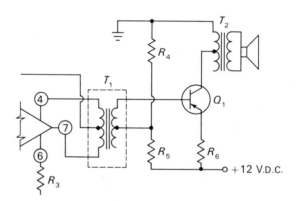

(b) Modification of basic IC amplifier for 4-watt,
class-A operation.

Figure 10.9
Circuit diagram of (a) basic IC amplifier and (b) modification
of basic IC amplifier for 4-watt, class-A operation.

R_5. 33-ohm, ½-watt.

T_1. Transistor transformer: primary, approximately 500 ohms
centertap; secondary, approximately 75 ohms (Argonne type
163, using CT and one side of winding for secondary, or equivalent.

T_2. Transistor output transformer: primary, approximately 24 ohms; secondary, to match load impedance (Agonne type 172, using CT and one side of winding for primary, or equivalent). Q_1. 2N2148.

Functional Description

The CA 3020 IC is a multistage, direct-coupled amplifier consisting of seven transistors and three diodes and resistors associated with the transistor biasing and temperature stabilization. The entire IC is packaged in a case approximately 0.3 inches in diameter. The resistors, capacitors, and transformer added externally to the IC form the input-and-output impedance-matching networks. A centertap transformer, T_1, is used to match the 120-ohm push-pull output to a standard 4- or 8-ohm loudspeaker load.

When the IC is used as a driver for a transistor output stage, a different output-stage biasing resistor, R_3, and a different matching transformer are used to couple the IC output to the input of the power transistor. The power transistor, Q_1, is biased by resistors R_4, R_5, and R_6; Q_1 is coupled to an 8-ohm loudspeaker load through transformer T_2.

Construction Notes

The CA 3020 should have a heat sink added to its case if the amplifier is to be used for more than 100 milliwatts of power. A snap-on type heat sink for use with a TO-5 transistor is adequate (International Rectifier type DD-121-C semiconductor heat sink, or equivalent). In addition to the heat sink, adequate ventilation should be provided when the IC is being used for high power. Maximum IC case temperatures of 150° C should not be exceeded.

When constructing the 4-watt output stage, use a heat sink for the 2N2148 transistor. A suitable heat sink may be constructed from a piece of aluminum approximately 4 by 2 by ⅛ inches.

The use of an IC socket (Augat type 8058-1G51 IC socket, or equivalent) is recommended for the CA 3020. If a socket is not used, care should be exercised in soldering the IC leads. In either case, the leads should not be shortened, or insertion will be difficult.*

---

* For detailed descriptions of the CA 3020 IC, along with applications and specifications, see both *RCA Semiconductor Products Databook* SPD-100, file no. 238, and *RCA Application Note* ICAN-5320, "Application of the RCA CA 3020 Multipurpose Integrated Circuit Audio Power Amplifier."

Appendixes

APPENDIX A
Schematic Symbols Used in Circuit Diagrams

Resistor	
Potentiometer	
Tapped resistor	
Rheostat	or
Fixed capacitor	
Fixed capacitor, electrolytic	
Adjustable or variable capacitor	
Switch, SPST	
Switch, SPDT	
Switch, DPST	
Switch, DPDT	
Switch, rotary	
Transistor, NPN	
Transistor, PNP	
Transistor, Field Effect	

Transistor, Unijunction

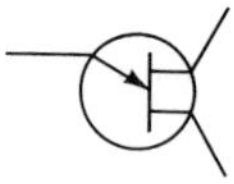

Diode, rectifier 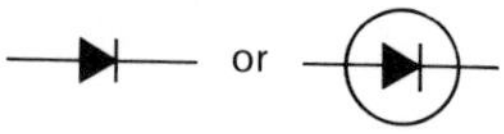   or

Diode, Zener   or

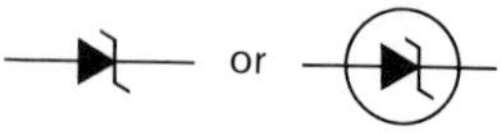

Diode, snap or bistable

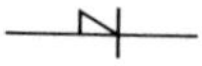

Diode, surge suppressor

Diode, tunnel   or

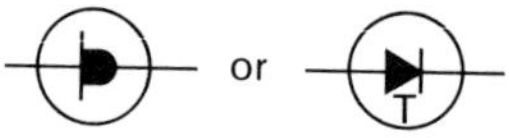

Diode, photo

Light-activated switch 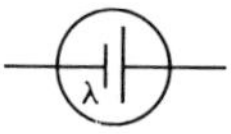   Photovoltaic    Photoresistive

Silicon-controlled rectifier   or

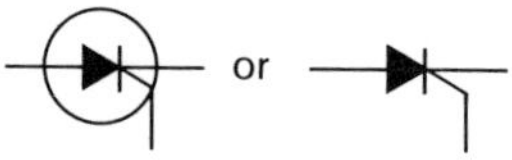

Relay, SPDT

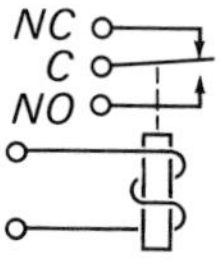

NC
C
NO

Relay, stepper

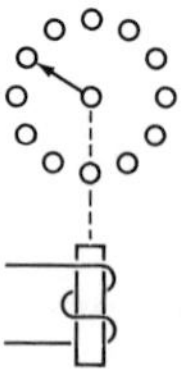

Relay, sequencing

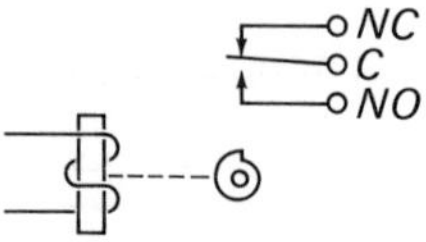

NC
C
NO

Air-core choke coil

Iron-core choke coil

Audio-frequency transformer (iron-core)

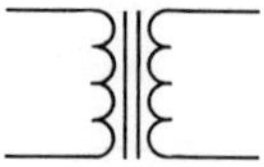

Radio-frequency transformer (air-core)

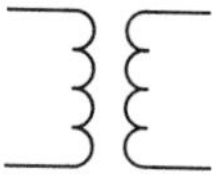

Power transformer

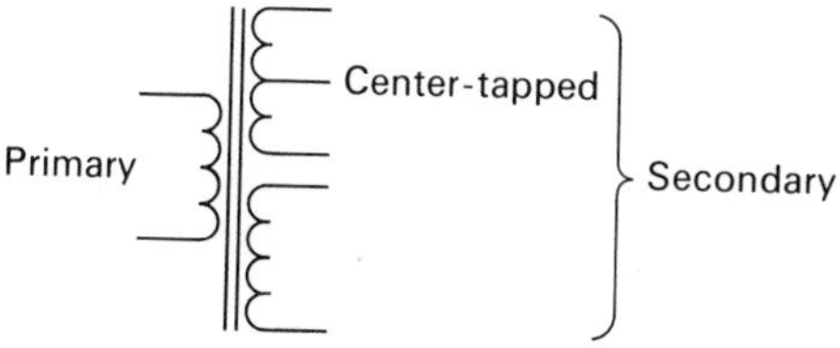

Primary

Center-tapped

Secondary

Pilot lamp

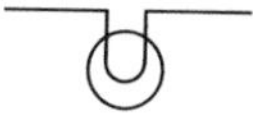

Loudspeaker

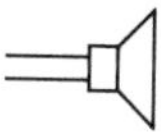

Headphones

Antenna

Ground

Vacuum tube heater or filament

Vacuum tube cathode

Vacuum tube grid

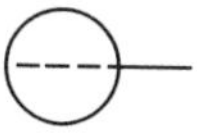

Vacuum tube plate

Three-element vacuum tube (triode)

Battery, one cell

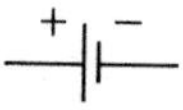

Battery, multicell

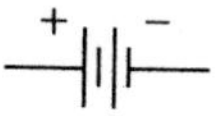

Fuse

APPENDIX B
Resistor and Capacitor Color Codes

RESISTOR COLOR CODES

Color	First digit A	Second digit B	Multiplier C	Tolerance D
Black	0	0	1	—
Brown	1	1	10	—
Red	2	2	100	—
Orange	3	3	1,000	—
Yellow	4	4	10,000	—
Green	5	5	100,000	—
Blue	6	6	1,000,000	—
Violet	7	7	10,000,000	—
Gray	8	8	100,000,000	—
White	9	9	—	—
Gold	—	—	0.1	±5%
Silver	—	—	0.01	±10%
No Color	—	—	—	±20%

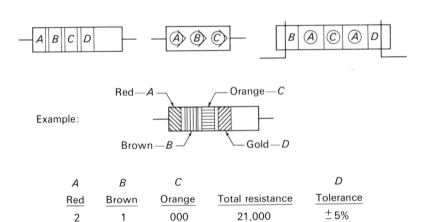

Example:

Red—A   Orange—C

Brown—B   Gold—D

A	B	C	D	
Red	Brown	Orange	Total resistance	Tolerance
2	1	000	21,000	±5%

MOLDED PAPER CAPACITOR CODES

Color	Digits of capacitance (mf) A	B	Multiplier C	Tolerance D	Tubular voltage rating (V. D.C.) E
Black	0	0	1	20%	—
Brown	1	1	10	—	100
Red	2	2	100	—	200
Orange	3	3	1,000	—	300
Yellow	4	4	10,000	—	400
Green	5	5	100,000	5%	500
Blue	6	6	1,000,000	—	600
Violet	7	7	—	—	700
Gray	8	8	—	—	800
White	9	9	—	10%	900
Gold	—	—	—	5%	1,000
Silver	—	—	—	10%	—
No Color	—	—	—	20%	—

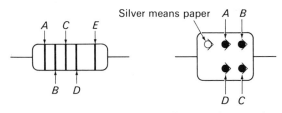

Tubular capacitor   Rectangular capacitors

MICA CAPACITOR CODES

Color	Digits of capacitance (mf) A	B	Multiplier C	Tolerance D
Black	0	0	1	±20%
Brown	1	1	10	—
Red	2	2	100	±2%
Orange	3	3	1,000	±3%
Yellow	4	4	10,000	—
Green	5	5	—	±5%
Blue	6	6	—	—
Violet	7	7	—	—
Gray	8	8	—	—
White	9	9	—	—
Gold	—	—	0.1	—
Silver	—	—	0.01	±10%

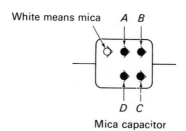

White means mica   A B

D C

Mica capacitor

APPENDIX C
Common Electronic Formulas

E—voltage (volts)
I—current (amperes)
L—inductance (henries)
R—resistance (ohms)
X_L—inductive reactance (ohms)
Z—impedance (ohms)

C—capacitance (farads)
h_z—frequency (hertz)
P—power (watts)
t—time (seconds)
X_C—capacitive reactance (ohms)

DIRECT CURRENT

1. Ohm's law

$$E = IR = \frac{P}{I} = \sqrt{RP}$$

$$I = \frac{E}{R} = \frac{P}{E} = \sqrt{\frac{P}{R}}$$

$$R = \frac{E}{I} = \frac{P}{I^2} = \frac{E^2}{P}$$

$$P = EI = I^2R = \frac{E^2}{R}$$

2. Series circuits

$$R_{\text{total}} = R_1 + R_2 + \cdots + R_n$$
$$E_{\text{total}} = E_1 + E_2 + \cdots + E_n$$
$$I_{\text{total}} = I_1 = I_2 = \cdots = I_n$$
$$P_{\text{total}} = P_1 + P_2 = \cdots + P_n$$

3. Parallel circuits

$$\frac{1}{R_{\text{total}}} = \frac{1}{R_1} + \frac{1}{R_2} + \cdots + \frac{1}{R_n}$$

$$R_{\text{total}} = \frac{1}{\dfrac{1}{R_1} + \dfrac{1}{R_2} + \cdots + \dfrac{1}{R_n}}$$

$$E_{\text{total}} = E_1 = E_2 = \cdots = E_n$$

$$I_{\text{total}} = I_1 + I_2 + \cdots + I_n$$

$$P_{\text{total}} = P_1 + P_2 + \cdots + P_n$$

ALTERNATING CURRENT

1. Power

$$P_{\text{peak}} = E_{\max} I_{\max}$$

$$P_{\substack{\text{average} \\ \text{(effective)}}} = \frac{E_{\max} I_{\max}}{2} = \frac{E_{\max}}{\sqrt{2}} = \frac{I_{\max}}{\sqrt{2}}$$

2. Voltage

$$E_{\substack{\text{effective} \\ \text{(RMS)}}} = \frac{E_{\max}}{\sqrt{2}} = 0.707\ E_{\max}$$

$$E_{\max} = E_{\text{eff(RMS)}} \times \sqrt{2} = 1.414\ E_{\text{eff}}$$

3. Current

$$I_{\substack{\text{effective} \\ \text{(RMS)}}} = \frac{I_{\max}}{\sqrt{2}} = 0.707\ I_{\max}$$

$$I_{\max} = I_{\text{eff(RMS)}} \times \sqrt{2} = 1.414\ I_{\text{eff}}$$

INDUCTORS

1. Coil inductance

$$L = \frac{X_L}{2\pi h_z}$$

2. Inductive reactance

$$X_L = 2\pi h_z L$$

3. Inductors in series (No flux linkage between coils)

$$L_{\text{total}} = L_1 + L_2 + \cdots + L_n$$

4. Inductors in parallel

$$L_{\text{total}} = \frac{1}{\dfrac{1}{L_1} + \dfrac{1}{L_2} + \cdots + \dfrac{1}{L_n}}$$

CAPACITORS

1. Capacitance

$$C = \frac{1}{2\pi h_z X_c}$$

2. Capacitive reactance

$$X_c = \frac{1}{2\pi h_z C}$$

3. Capacitors in series

$$C_{\text{total}} = \frac{1}{\dfrac{1}{C_1} + \dfrac{1}{C_2} + \cdots + \dfrac{1}{C_n}}$$

4. Capacitors in parallel

$$C_{\text{total}} = C_1 + C_2 + \cdots + C_n$$

RESONANT CIRCUITS (AT RESONANCE)

$$X_L = X_c$$

$$2\pi h_z L = \frac{1}{2\pi h_z C}$$

$$h_z = \frac{1}{2\pi\sqrt{LC}}$$

CIRCUIT TIME CONSTANTS

$$t = RC$$

$$t = \frac{L}{R}$$

APPENDIX D

Table of Standard Annealed Bare Copper Wire
Using American Wire Gauge (B & S)

Gauge (AWG) or (B & S)	Diameter in inches	Area in circular mils	Resistance at 68° F Ohms per M'	Resistance at 68° F Feet per ohm	Current capacity (amps)— rubber insulated
0000	.4600	211600.	.04901	20400.	225
000	.4096	167800.	.06180	16180.	175
00	.3648	133100.	.07793	12830.	150
0	.3249	105500.	.09827	10180.	125
1	.2893	83690.	.1239	8070.	100
2	.2576	66370.	.1563	6400.	90
3	.2294	52640.	.1970	5075.	80
4	.2043	41740.	.2485	4025.	70
5	.1819	33100.	.3133	3192.	55
6	.1620	26250.	.3951	2531.	50
7	.1443	20820.	.4982	2007.	
8	.1285	16510.	.6282	1592.	35
9	.1144	13090.	.7921	1262.	
10	.1019	10380.	.9989	1001.	25
11	.09074	8234.	1.260	794.	
12	.08081	6530.	1.588	629.6	20
13	.07196	5178.	2.003	499.3	
14	.06408	4107.	2.525	396.0	15
15	.05707	3257.	3.184	314.0	
16	.05082	2583.	4.016	249.0	6
17	.04526	2048.	5.064	197.5	
18	.04030	1624.	6.385	156.5	3
19	.03589	1288.	8.051	124.2	
20	.03196	1022.	10.15	98.5	
21	.02846	810.1	12.80	78.11	
22	.02535	642.4	16.14	61.95	
23	.02257	509.5	20.36	49.13	
24	.02010	404.0	25.67	38.96	
25	.01790	320.4	32.37	30.90	
26	.01594	254.1	40.81	24.50	

Gauge (AWG) or (B & S)	Diameter in inches	Area in circular mils	Resistance at 68° F		Current capacity (amps)— rubber insulated
			Ohms per M′	Feet per ohm	
27	.01420	201.5	51.47	19.43	
28	.01264	159.8	64.90	15.41	
29	.01126	126.7	81.83	12.22	
30	.01003	100.5	103.2	9.691	
31	.008928	79.7	130.1	7.685	
32	.007950	63.21	164.1	6.095	
33	.007080	50.13	206.9	4.833	
34	.006305	39.75	260.9	3.833	
35	.005615	31.52	329.0	3.040	
36	.005000	25.00	414.8	2.411	
37	.004453	19.83	523.1	1.912	
38	.003965	15.72	659.6	1.516	
39	.003531	12.47	831.8	1.202	
40	.003145	9.888	1049.	0.9534	
41	.00280	7.8400	1323.	.7559	
42	.00249	6.2001	1673.	.5977	
43	.00222	4.9284	2104.	.4753	
44	.00197	3.8809	2672.	.3743	
45	.00176	3.0976	3348.	.2987	
46	.00157	2.4649	4207.	.2377	

APPENDIX E
Basic Relay Circuits

The following list constitutes a guide to useful relay circuit configurations. It does not specify components or operating voltages, since it is intended to present general operating descriptions from which useful circuits may be constructed.

An inductive load such as a relay or solenoid can produce several hundred volts by self-induction when its circuit is suddenly interrupted. This high-voltage

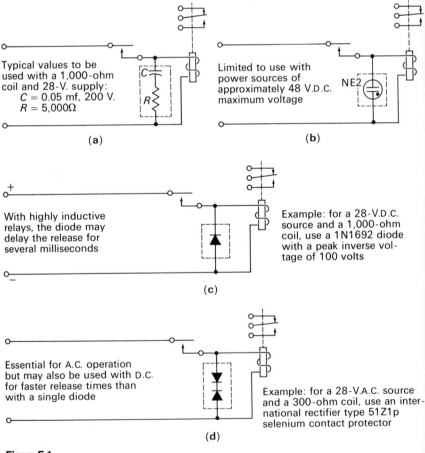

Typical values to be used with a 1,000-ohm coil and 28-V. supply:
$C = 0.05$ mf, 200 V.
$R = 5,000\Omega$

(a)

Limited to use with power sources of approximately 48 V.D.C. maximum voltage

(b)

With highly inductive relays, the diode may delay the release for several milliseconds

Example: for a 28-V.D.C. source and a 1,000-ohm coil, use a 1N1692 diode with a peak inverse voltage of 100 volts

(c)

Essential for A.C. operation but may also be used with D.C. for faster release times than with a single diode

Example: for a 28-V.A.C. source and a 300-ohm coil, use an international rectifier type 51Z1p selenium contact protector

(d)

Figure E.1
Typical spark-suppression networks: **(a)** *R-C* network; **(b)** neon bulb; **(c)** single diode; **(d)** dual diode.

surge can cause sparking and consequent failure and erosion of contacts used to break the load, and can also cause spurious signals to pass to other pieces of electronic equipment. For these reasons, suitable spark-suppression networks should be incorporated in completed relay circuits to protect the contacts that break inductive loads. Although spark-suppression networks have been left out of the basic relay circuits given below, typical networks are described in Figure E-1.

With equipment that operates at relay switching speeds, any of the spark suppressors in Figure E-1 should work adequately. For use with high-speed, solid-state switching circuits, the R-C circuit provides the most rapid and efficient arc suppression. Precise values of R and C may be established empirically, using an oscilloscope to observe the amount of suppression. Other transient-voltage suppressors include varistors and thyrectors, both solid-state devices especially made for high-speed transient-voltage suppression and available in a variety of specified threshold voltages.

ACCELERATED ACTIVATION

Accelerated activating time may be achieved by using supply voltages that are higher than the relay operating voltage (see Figure E-2). The high supply

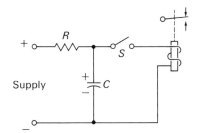

Figure E.2
Accelerated actuation.

voltage is used to charge a capacitor through a current limiter, R. When switch S is closed, the capacitor will discharge through the relay coil. The momentary overvoltage applied to the relay coil remains while the capacitor discharges to the

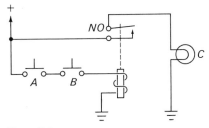

Figure E.3
AND gate.

relay holding voltage that is established by the voltage drop across R. Supply voltages of up to ten times the relay voltage may be used, resulting in operating times shortened by a factor of approximately five. The relay life will be shortened by approximately the same factor.

AND GATE

See Figure E-3, and see page 160 for a detailed description.

ADJUSTABLE CIRCUIT BREAKER

As shown in Figure E-4, relay K senses the voltage developed across adjustable resistor R_1, which is in series with the load. When the load current is too high,

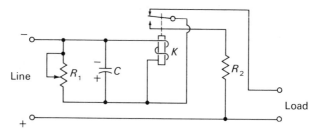

Figure E.4
Adjustable circuit breaker.

enough voltage is developed across R_1 to activate K. Capacitor C assures reliable operation of the relay by maintaining relay operating current after the load is disconnected through the relay contacts. R_2 provides a safe holding current for the relay contacts. R_2 provides a safe holding current for the relay, once the load is disconnected.

LATCHING-DOWN OPERATION

In the circuit shown in Figure E-5, when contact A is closed, the relay becomes inoperative and remains inoperative until A is open and S is closed. Resistor R prevents a short circuit through the shunt path.

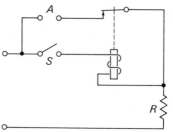

Figure E.5
Latching-down operation.

LATCHING-UP OPERATION

In the circuit shown in Figure E-6, on momentary closure of contact A, the relay energizes and remains energized until push-button switch S is opened.

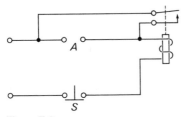

Figure E.6
Latching-up operation.

MULTIVIBRATOR

In the circuit shown in Figure E-7, on application of power, K_1 and K_2 will attempt to activate simultaneously, but slight differences in their construction will

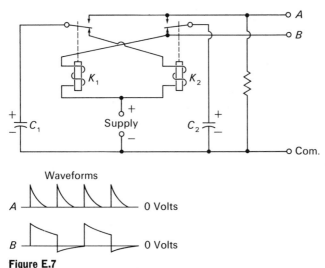

Figure E.7
Multivibrator.

cause one to activate first. Assuming K_1 activates first, it will remove operating power from K_2 and remain activated until capacitor C_2 is charged and operating current can no longer flow through K_1. When K_1 releases, capacitor C is placed in series with the battery and K_2; K_2 activates and remains activated until C_1 is charged and operating current can no longer flow. With K_2 activated, C_2 is able to discharge through R. The wave form across R is a series of voltage spikes, while the wave form across K_1 approximates a square wave at one half the frequency of the voltage spikes.

NAND GATE

See Figure E-8, and see page 160 for a detailed description.

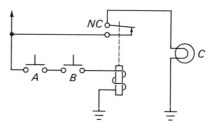

Figure E.8
NAND gate.

NOR GATE

See Figure E-9, and see page 160 for a detailed description.

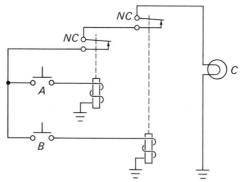

Figure E.9
NOR gate.

OR GATE

See Figure E-10, and see page 160 for a detailed description.

POLARIZED RELAY OPERATION

In the circuit shown in Figure E-11, the bridge rectifier provides a constant polarity to the relay coil irrespective of the operating voltage polarity.

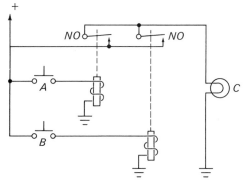

Figure E.10
OR gate.

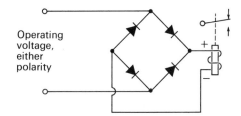

Figure E.11
Polarized relay operation.

PULSE FORMER

In the circuit shown in Figure E-12, on application of power, capacitor C_1 is charged through R and K_1. When a ground pulse is delivered to the "operate" terminal, K_1 activates and C_1 discharges. K_2 operates through the *NO* K_1 contacts. K_1 remains energized until C_1 again charges through R and K_1; K_2 will

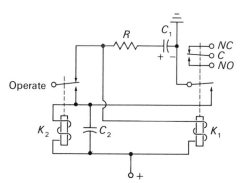

Figure E.12
Pulse former.

remain latched through the operate terminal and K_2 contacts, if the operate pulse is longer than the K_1 operate time, or through the *NO* K_1 contacts, if the operate pulse is shorter than the K_1 operate time. In either case, K_1 will remain activated until C_1 is charged sufficiently to lower K_1's operating voltage. C_2 (approximately 0.5 mf) insures reliable operation of K_2.

PULSER

See Figure E-13, and see p. 144 for a detailed circuit description.

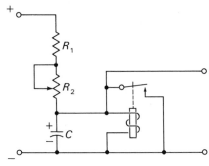

Figure E.13
Pulser.

PULSE STRETCHER AND FREQUENCY DIVIDER

In the circuit shown in Figure E-14, the trigger pulse must be of sufficient amplitude and duration to momentarily activate the relay. Pulse duration is established by R and C. The circuit can also function as a frequency divider by

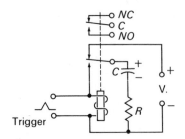

Figure E.14
Pulse stretcher and
frequency divider.

increasing the R-C time constant so the relay remains energized during several trigger pulses. For operation as a frequency divider, the amplitude, width, and frequency of the trigger pulse must be nearly constant.

RETARDED ACTIVATION—SLOW OPERATE, FAST RELEASE

In the circuit shown in Figure E-15, the capacitor, shunted across the relay coil, delays activation of the relay. When it is activated, the capacitor is removed from across the coil and is discharged through R_2. When S is opened, the relay drops out. Resistor R_1 limits surge currents to the capacitor and regulates the delay.

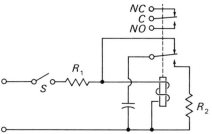

Figure E.15
Retarded operation—slow
operate, fast release.

RETARDED ACTIVATION—SLOW OPERATE, SLOW RELEASE

In the circuit shown in Figure E-16, the capacitor, shunted across the relay coil, delays both the activation and release of the relay; R limits surge currents to the capacitor and regulates the time delay.

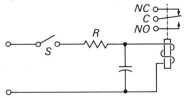

Figure E.16
Retarded operation—slow
operate, slow release.

SHUNT OPERATION

In the circuit shown in Figure E-17, the relay energizes when contact A is open and releases when contact A is closed; R prevents short circuits across the shunt network.

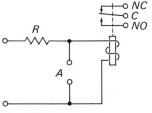

Figure E.17
Shunt operation.

SQUARE-WAVE GENERATOR

See Figure E-18, and see page 145 for a detailed circuit description.

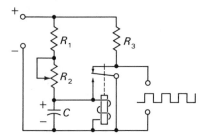

Figure E.18
Square-wave generator.

TRANSISTOR, RELAY AMPLIFIER

See Figure E-19, and see Chapter 1 for a detailed description of transistor amplifier principles.

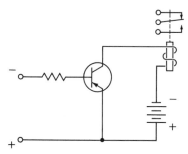

Figure E.19
Transistor, basic relay amplifier.

TRANSISTOR, BASIC TIMING RELAY

See Figure E-20, and see Chapter 9 for a detailed description of basic timing principles.

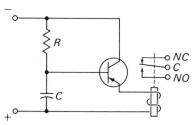

Figure E.20
Transistor, basic timing relay.

TRANSISTOR, RELAY LATCHING

In the circuit shown in Figure E-21, relay turn-on is similar to the basic latching relay. Turn-off is accomplished by momentarily cutting off the transistor.

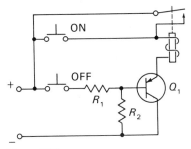

Figure E.21
Transistor, latching relay

VOLTMETER, BASIC

In the circuit shown in Figure E-22, the voltage to be measured is connected,
with R initially set to maximum resistance. As the resistance is reduced, voltage
across the relay will reach the firing point; R may be roughly calibrated to indicate
various values of input voltage required to operate the relay.

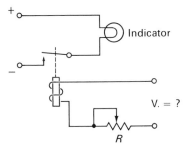

Figure E.22
Voltmeter, basic.

VOLTMETER, PRECISION

In the circuit shown in Figure E-23, relay K will not operate until the Zener voltage, V_1, is exceeded by the relay operating voltage, V_2. With R set to zero resistance, the unknown voltage must be equal to at least V_1 plus V_2 in order to

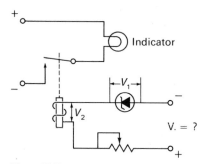

Figure E.23
Voltmeter, precision.

activate K. With R set to equal the coil resistance, K will activate when the unknown voltage is equal to at least V_1 plus $2V_2$. The value of R may be closely calibrated between the two voltage limits.

SELECTED BIBLIOGRAPHY

INTRODUCTION TO ELECTRICITY

Cornsweet, Tom N. *The Design of Electric Circuits in the Behavioral Sciences.* New York: John Wiley and Sons, 1965.

Mileaf, Harry, ed. *Electricity One-Seven.* New York: Hayden Book Co., 1967.

New York Institute of Technology. *A Programmed Course in Basic Electricity.* New York: McGraw-Hill Book Co., 1967.

INTRODUCTION TO ELECTRONICS

Kiver, Milton S. *Transistors in Radio, Television, and Electronics.* New York: McGraw-Hill Book Co., 1959.

New York Institute of Technology. *A Programmed Course in Basic Electronics.* New York: McGraw-Hill Book Co., 1967.

RCA Service Co. *Fundamentals of Transistors.* Englewood Cliffs, New Jersey: Prentice-Hall, Inc., 1967.

Squires, Terence L. *Beginner's Guide to Electronics.* New York: Philosophical Library, 1968.

Training Systems, Inc., and Stanley L. Levine. *Simplified Transistor Theory.* New York: Hayden Book Co., 1967.

CIRCUIT DESIGN AND CONSTRUCTION

Benedict, R. Ralph. *Electronics for Scientists and Engineers.* Englewood Cliffs, New Jersey: Prentice-Hall, Inc., 1967.

Federal Electric Corporation. *Math for Electronics.* Englewood Cliffs, New Jersey: Prentice-Hall, Inc., 1967.

Patrick, Owen G. *Creative Electronics Fabrication.* New York: Holt, Rinehart, and Winston, 1968.

Riddle, Robert L., and Marlin P. Ristenbatt. *Transistor Physics and Circuits.* Englewood Cliffs, New Jersey: Prentice-Hall, Inc., 1958.

Ritchie, George L. *Electronics Construction Techniques.* New York: Holt, Rinehart, and Winston, 1967.

Texas Instruments Incorporated, Engineering Staff. *Transistor Circuit Design.* New York: McGraw-Hill Book Co., 1963.

TROUBLESHOOTING AND TESTING

Darr, Jack. *How to Test Almost Everything Electronic.* New York: Gernsback Library, 1966.

Middleton, Robert G. *Troubleshooting with the Oscilloscope*. New York: Howard W. Sams and Co., 1964.

Schuster, Donald H. *Logical Electronic Troubleshooting*. New York: McGraw-Hill Book Co., 1967.

SENSING AND CONTROL CIRCUITS

Appels, J. Thomas, and B. H. Geels. *Handbook of Relay Switching Technique*. New York: Philips Technical Library, 1966.

Giles, A. F. *Electronic Sensing Devices*. Cleveland, Ohio: Chemical Rubber Co., 1968.

Markus, John, and V. Zeluff. *Handbook of Industrial Control Circuits*. New York: McGraw-Hill Book Co., 1956.

Shields, John Potter. *How to Build Proximity Detectors and Metal Locators*. New York: Howard W. Sams and Co., 1965.

Turner, Rufus P. *Bridges and Other Null Devices*. New York: Howard W. Sams and Co., 1967.

PROGRAMMING CIRCUITS

Maley, Gerald A., and John Earle. *The Logic Design of Transistor Digital Computers*. Englewood Cliffs, New Jersey: Prentice-Hall, Inc., 1963.

Oppenheimer, Samuel L. *Semiconductor Logic and Switching Circuits*. Columbus, Ohio: Charles E. Merrill Books, 1966.

Richards, R. K. *Electronic Digital Components and Circuits*. Princeton, New Jersey: D. Van Nostrand Co., 1967.

Scientific American. *Information*. San Francisco: W. H. Freeman and Co., 1966.

USEFUL REFERENCE MANUALS

Allied Electronics. *Data Handbook*. Chicago: Allied Radio Corporation, 5th ed., 1968.

General Electric. *Controlled Rectifier Manual*. Syracuse, New York: General Electric Company, Semiconductor Products Department, 1960.

———. *Transistor Manual*. Syracuse, New York: General Electric Company, Semiconductor Products Department, 1969.

RCA. *Transistor Manual SC-13*. Harrison, New Jersey: Radio Corporation of America, Electronics Components and Devices Division, 1969.

MONTHLY MAGAZINES

Electronics Illustrated.
Electronics World.
Popular Electronics.
Radio-Electronics.

SUPPLIERS' CATALOGUES

Allied Radio. 100 North Western Avenue, Chicago, Illinois.

Brill Electronics Corporation. 233 West Madison Street, Chicago, Illinois.

Burstein-Applebee Company. 1012 McGee Street, Kansas City, Mo.

Herback and Rodeman, Inc. 1204 Arch Street, Philadelphia, Penn.

Lafayette Radio Electronics. 111 Jericho Turnpike, Syosset, Long Island, New York.

Newark Electronics Corporation. 233 West Madison Street, Chicago, Illinois.

Universal Relay Corporation. 42 White Street, New York, New York.

Index